The Subregular Germ of Orbital Integrals

Recent Titles in This Series

(Continued in the back of this publication)

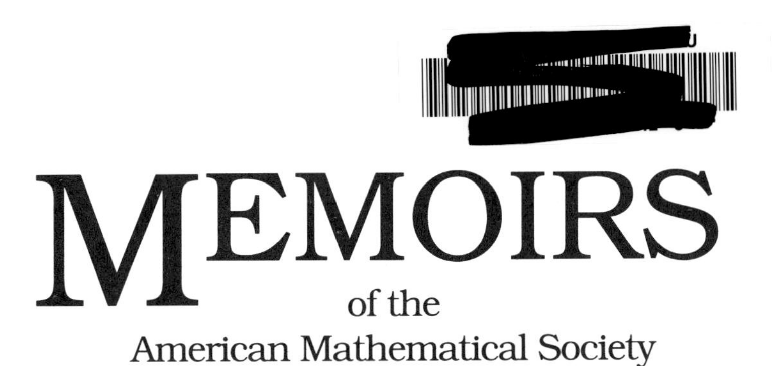

Number 476

The Subregular Germ of Orbital Integrals

Thomas C. Hales

September 1992 • Volume 99 • Number 476 (third of 4 numbers) • ISSN 0065-9266

American Mathematical Society
Providence, Rhode Island

1991 *Mathematics Subject Classification.*
Primary 20G25; Secondary 22E35, 12B27.

Library of Congress Cataloging-in-Publication Data

Hales, Thomas Callister.
The subregular germ of orbital integrals/Thomas C. Hales.
p. cm. – (Memoirs of the American Mathematical Society, ISSN 0065-9266; no. 476)
"September 1992, volume 99, number 476 (third of 4 numbers)."
Includes bibliographical references (p.).
ISBN 0-8218-2539-9
1. p-adic fields. 2. Representations of groups. 3. Germs (Mathematics) I. Title. II. Title: Orbital intervals. III. Series.
QA3.A57 no. 476
[QA247]
510 s–dc20 92-18060
[512′.74] CIP

Memoirs of the American Mathematical Society

This journal is devoted entirely to research in pure and applied mathematics.

Subscription information. The 1992 subscription begins with Number 459 and consists of six mailings, each containing one or more numbers. Subscription prices for 1992 are $292 list, $234 institutional member. A late charge of 10% of the subscription price will be imposed on orders received from nonmembers after January 1 of the subscription year. Subscribers outside the United States and India must pay a postage surcharge of $25; subscribers in India must pay a postage surcharge of $43. Expedited delivery to destinations in North America $30; elsewhere $82. Each number may be ordered separately; *please specify number* when ordering an individual number. For prices and titles of recently released numbers, see the New Publications sections of the *Notices of the American Mathematical Society.*

Back number information. For back issues see the *AMS Catalog of Publications.*

Subscriptions and orders should be addressed to the American Mathematical Society, P. O. Box 1571, Annex Station, Providence, RI 02901-1571. *All orders must be accompanied by payment.* Other correspondence should be addressed to Box 6248, Providence, RI 02940-6248.

Memoirs of the American Mathematical Society is published bimonthly (each volume consisting usually of more than one number) by the American Mathematical Society at 201 Charles Street, Providence, RI 02904-2213. Second-class postage paid at Providence, Rhode Island. Postmaster: Send address changes to Memoirs, American Mathematical Society, P. O. Box 6248, Providence, RI 02940-6248.

Printed in the United States of America.
The paper used in this book is acid-free and falls within the guidelines
established to ensure permanence and durability. ♾

10 9 8 7 6 5 4 3 2 1 97 96 95 94 93 92

Contents

ABSTRACT

An integral formula for the subregular germ of a κ-orbital integral is developed. The formula holds for any reductive group over a p-adic field of characteristic zero. This expression of the subregular germ is obtained by applying Igusa's theory of asymptotic expansions. The integral formula is applied to the question of the transfer of a κ-orbital integral to an endoscopic group. It is shown that the quadratic characters arising in the subregular germs are compatible with the transfer. Details of the transfer are given for the subregular germ of unitary groups.

Key words and phrases. p-adic group, orbital integral, Shalika germs, Igusa theory, Langlands theory, representation theory, unipotent classes, Dynkin curve.

INTRODUCTION

An elementary fact in the theory of finite groups states that the vector space spanned by irreducible characters of representations of a group coincides with the vector space spanned by characteristic functions of conjugacy classes. To give an explicit formula for an irreducible character is to express a character as a linear combination of the basis vectors formed by characteristic functions. For Lie groups or p-adic groups, irreducible characters must be interpreted as distributions. Similarly the characteristic functions must be replaced by distributions, orbital integrals, supported on the conjugacy classes if one hopes to develop systematically a theory of characters on Lie and p-adic groups. A successful character theory of these groups should give relations expressing distribution characters as linear combinations of orbital integrals, and orbital integrals as sums of characters. This work is concerned with the study of orbital integrals on p-adic groups, needed for eventual applications to automorphic representation theory and the trace formula.

These orbital integrals have a notoriously complicated structure. As the conjugacy class is allowed to vary, the orbital integrals possess an asymptotic expansion called the Shalika germ expansion. In contrast to what the terminology might suggest, the asymptotic expansion has only finitely many terms and for p-adic groups actually gives an exact formula for the orbital integral in a sufficiently small neighborhood of the identity element. Moreover, by inductive arguments the behavior of an orbital integral may be understood once its behavior near the identity element of the group is understood. Consequently most questions we might have about orbital integrals can be answered from Shalika's expansion. Unfortunately, Shalika's *existence* proof of an asymptotic expansion has not resulted in explicit formulas for the germs except in a few elementary cases.

A basic problem of harmonic analysis on reductive p-adic groups is then to develop expressions for the terms of the Shalika expansion of orbital integrals. This work uses a geometrical approach, introduced by Langlands and Shelstad, to calculate the first two terms of the Shalika expansion. These terms are called the regular and subregular terms of the expansion. The first term of the expansion is, with suitable normalizations of measures, an invariant integral over the stable regular unipotent class. The second term of the expansion, as we will see, is a sum of integrals of the form

$$\theta(\lambda)|\lambda| \int_{\mathbb{P}^1} \frac{dv}{|v|} \int_{\mathbb{P}^1} \eta(p(w)) \frac{dw}{|w|^2} \mu_0$$

where p is an appropriate polynomial in w; and η, θ are multiplicative characters on the p-adic field F. Also μ_0 is an invariant integral over a subregular unipotent conjugacy class of the group. These integrals over projective lines must be understood as principal value integrals.

The explicit formulas for the first two terms of the Shalika expansion allow us to check a number of conjectures concerning orbital integrals – many of which were known previously only for a few groups of small rank. This method used here to obtain explicit formulas for Shalika germs remains the only known general method to obtain such formulas. Other methods are now available for the general and special linear groups.

Here are a few remarks on the geometrical construction of Shalika germs. If Γ is a curve inside a Cartan subgroup T of G such that $\Gamma(0) = 1 \in T$ and $\Gamma(s)$, $s \neq 0$ is a regular element of T, then the conjugacy class of $\Gamma(s)$, $s \neq 0$ may be identified with $\Gamma(s) \times T\backslash G$ and the Shalika germ expansion gives the asymptotics of integrals on $\Gamma(s) \times T\backslash G$ as s tends to 1. The essential part of the construction of Shalika germs is the construction of a G-equivariant completion $Y \to^{\pi} \Gamma$ of $\Gamma \times T\backslash G \to^{\pi} \Gamma$. The theory of Igusa then states that the asymptotic expansion as s tends to 1 may be understood by studying the divisor $D = \pi^{-1}(\Gamma(1))$ in the variety Y. Roughly each term of the asymptotic expansion is an integral over some of the irreducible components of D.

To see some of the technical difficulties involved in this procedure, consider two completions Y_1, Y_2 of $\Gamma \times T\backslash G$ with corresponding divisors D_1 and D_2. The fact that the asymptotic expansion is independent of the completion shows that the terms of the expansion (integrals over components of D_1 and D_2) coincide. This suggests that these integrals on D_1 and D_2 should be invariant under a large class of birational maps. This is indeed the case. In the special case that Y_1 is obtained from Y_2 by blowing up Y_1 along a subvariety of D_1, it suggests that the exceptional divisors introduced by blowing up usually make no contribution to the asymptotic expansion. In other words, if we begin with a completion Y_1 which is far from being a minimal completion, it will be necessary to sort through a large number of exceptional or spurious irreducible components of D_1 that ultimately make no contribution to the asymptotic expansion. Unfortunately, we know of no better general completion of $\Gamma \times T\backslash G$ than the one introduced below, and a great deal of work is needed to eliminate all but a few *fundamental* components of D_1 that lead to the Shalika expansion. Finally we should remark that the variety Y is singular, so that it is necessary to prove that the singularities are of such a nature that they do not affect the Shalika germ expansion.

Langlands has conjectured deep relations between integrals and representations of p-adic on different groups (the theory of endoscopy). Some of these are formulated as or translate into conjectural identities between orbital integrals on different groups. We may at first hope for these identities to hold for geometrical reasons. To explain this idea, suppose that $\int_{X_1} f_1 d\omega_1$ is an integral formula for a Shalika germ on G and $\int_{X_2} f_2 d\omega_2$ is an integral formula for a Shalika germ on H. Suppose that we are to show that these two integrals are equal. We might hope for a birational map $\phi : X_1 \to X_2$ carrying $d\omega_1$ to $d\omega_2$ and f_1 to f_2. If such a birational map satisfied certain technical hypotheses, we could then conclude that the two integrals are equal. In the cases that have been worked out in detail, this expectation has been fulfilled in a slightly weaker form. There have been geometrical decompositions of the varieties X_1, X_2 such that by geometrical "cut and paste" operations the identities $\int_{X_1} f_1\omega_1 = \int_{X_2} f_2\omega_2$ were established. Thus these identities are established without computing any integrals. This I take to be one of the strengths of the theory developed here.

Chapters I and II are preliminary. They describe a number of auxiliary varieties used to construct the varieties of ultimate concern to us. Some useful coordinates on these varieties are developed. Beginning in chapter III, we turn from general considerations to focus on the subregular unipotent classes and their germs. If Igusa theory is to be successfully applied a large number of divisors must be systematically excluded from consideration. Chapter III shows how to exclude these spurious divisors for groups of rank two and chapter IV excludes them for groups of higher rank. In chapters V and VI we give explicit formulas for the data entering into the integral representation of the subregular germ. We show for instance that the irreducible components of the surface (giving the explicit formula for the subregular germ) are in bijection with the lines of the Dynkin curve and that each irreducible component is a rational surface. Chapter VII discusses applications to the transfer of κ-orbital integrals to endoscopic groups. A stable subregular germ is shown to be equal up to a constant to a stable subregular germ on a quasi-split inner form. We show that the quadratic characters θ arising in the subregular germ are compatible with the transfer. Finally we give the details of the transfer of the subregular germ for unitary groups.

This work appeared originally as the author's thesis under the direction of Professor R.P. Langlands. I would like to thank R.P. Langlands for introducing me to this fruitful field of thought and for his continued encouragement and assistance.

I. BASIC CONSTRUCTIONS

Chapter I is concerned with preliminary constructions. The groups are not assumed to be quasi-split. The groups are taken over a p-adic field F of characteristic zero. The germs are studied near the group identity. We are not always careful in distinguishing a group G from its elements over $\bar{F}$ so that expressions such as $g \in G$ should be interpreted as $g \in G(\bar{F})$.

1. Background Information.

Igusa has introduced a method of studying asymptotic expansions of integrals over a local field. The expansion holds in the following context. A variety is fibred over a punctured neighborhood of a point p on a curve. Let λ be a local parameter at p. The integral is taken over a fibre and consequently depends on the parameter λ. Igusa theory gives, provided a number of technical conditions are satisfied, an asymptotic expansion of the integral as λ tends to 0. The theory gives explicit formulas for the coefficients of the asymptotic expansion. The locus of $\lambda = 0$ in the variety is to be a union of divisors. The coefficients of the asymptotic expansion are given as principal value integrals over the divisors.

R.P. Langlands [15] applies Igusa theory to the study of κ-orbital integrals by constructing a variety and a curve such that the integral taken over the fibre of the curve is equal to a κ-orbital integral. Chapter I is devoted to a study of the variety he constructs. Some useful coordinates are defined that simplify computations in the variety.

2. The Igusa Variety.

This section reviews the construction of the variety Y_1 introduced in [15]. The variety Y_1 and its resolution Y_Γ are constructed using a number of auxiliary varieties. First I will make a list of these varieties for reference, and then give the definitions.

S^0 is the variety of regular stars

S is the variety of stars, the closure of S^0

S' is the subvariety of S such that for each simple root α there is at least one chamber $W(\omega)$ for which $z(W(\omega), \alpha) \neq 0$

$S_1(B_\infty, B_0)$ is the fibred product $S_1(B_\infty, B_0) = S'(B_\infty, B_0) \times_{T_0} \mathbb{A}^r$.

S_1 is a first resolution of S

S'' is the open subvariety of S_1 given on each open patch $S''(B_\infty, B_0)$ by

$$S^0(B_\infty, B_0) \times_{T_0} \mathbb{A}^r \subseteq S'(B_\infty, B_0) \times_{T_0} \mathbb{A}^r = S_1(B_\infty, B_0).$$

Received by the editor February 20, 1989

X^0 is a subvariety of $G \times S^0$
X is the closure of X^0 in $G \times S$
X' is the closure of X^0 in $G \times S'$
X_1 is the closure of X^0 in $G \times S_1$
X'' is the closure of X^0 in $G \times S''$
Y^0 is the restriction of X^0 to the inverse image of a curve Γ in T
Y is the closure of Y^0 in X
Y' is the closure of Y^0 in X'
Y_1 is the closure of Y^0 in X_1
Y'' is the closure of Y^0 in X''
Y_Γ is a G-equivariant resolution of Y_1 which satisfies the conditions of Igusa data.

Let G be a reductive group defined over a p-adic field F of characteristic zero. Let $T \subseteq G$ be a Cartan subgroup defined over F and fix a Borel subgroup $\mathbb{B}$ containing T (which need not be defined over F). Let Ω be the Weyl group of G with respect to T. Let W_+ be the positive Weyl chamber with respect to $\mathbb{B}$ and let $W(\omega)$ denote the Weyl chamber $\omega^{-1}W_+$. Then the Borel subgroups containing T may be indexed by the Weyl chambers by setting $\mathbb{B}^\omega = \mathbb{B}(W)$ where $W = W(\omega)$.

Consider the n-fold product of the variety of Borel subgroups V^n where $n = |\Omega|$. The group G acts on V^n by $(B_1, \ldots, B_n).g = (B_1^g, \ldots, B_n^g)$. The variety of *regular stars* S^0 is defined to be the G-orbit of the point $(\mathbb{B}(W))$ in V^n. The variety of *stars* S is the closure of S^0 and is a projective variety. Let T^0 be the set of regular elements of T. There is a morphism from $T^0 \times T\backslash G$ to $G \times S^0$ given by $(t, g) \rightarrow (t^g, (\mathbb{B}(W)^g))$. Let X^0 denote the image of $T^0 \times T\backslash G$ in $G \times S^0$ under this morphism and let X be its closure in $G \times S$.

A morphism from X to T is defined as follows. If $(g, (B(W)))$ is a point in X, then select $h \in G(\bar{F})$ such that $B(W_+)^h = \mathbb{B}(W_+) = \mathbb{B}$. Then g^h lies in $\mathbb{B}$. Then (g^h modulo N) $\in \mathbb{B}/N \simeq T$ where N is the unipotent radical of $\mathbb{B}$. This map is independent of the choice of h. The composite $T^0 \times T\backslash G \rightarrow T$ equals the projection onto the first factor.

We introduce coordinate patches $S(B_\infty, B_0)$ of S and coordinates $z(W, \alpha)$ on $S(B_\infty, B_0)$ as follows. Let B_∞ and B_0 be opposite Borel subgroups with intersection T_0. For each simple root α fix root vectors X_α and $X_{-\alpha}$ for T_0 in the Lie algebra of G such that $[X_\alpha, X_{-\alpha}] = H_\alpha$ with $\alpha(H_\alpha) = 2$. Let $S(B_\infty)$ be the set of stars $(B(W))$ in S such that $B(W)$ is opposite B_∞ for all W. Consider one such point $(B(W))$ in S. Fix a Weyl chamber W_1 and simple root α. If $W_1 = W(\omega)$ then write $W_2 = W(\sigma_\alpha\omega)$. We can write $B(W_1) = B_0^{\nu_1}, B(W_2) = B_0^{\nu_2}$ with ν_1 and $\nu_2 \in N_\infty$ the unipotent radical of B_∞. The parabolic subgroup of type α containing $B(W_1)$ also contains $B(W_2)$ so that B_0 and $B_0^{\nu_2\nu_1^{-1}}$ are opposite B_∞ and lie in the parabolic subgroup P_α of type α containing B_0. Thus $\nu_2\nu_1^{-1} = \exp(z(W_1, \alpha)X_{-\alpha})$ for some uniquely determined value $z(W_1, \alpha)$. Also let $\nu \in N_\infty$ be defined by $B(W_+) = B_0^\nu$. The variables $(z(W, \alpha)) : \forall\ (W, \alpha)$ together with the coefficients of ν generate the coordinate ring of $S(B_\infty)$. Also let $S(B_\infty, B_0) = \{(B(W)) \subseteq S(B_\infty) : \nu = 1\}$. The varieties $X(B_\infty, B_0), Y(B_\infty, B_0)$, etc. have obvious definitions as subvarieties of X, Y, etc.

The pairs (W, α) and consequently the variables $z(W, \alpha)$ are in bijection with oriented walls of Weyl chambers. If W is a Weyl chamber and $\gamma = 0$ defines a wall of that chamber for some positive root γ, then there is an element of the Weyl group ω such that $W = W(\omega)$ and $\omega.\gamma$ is a simple root α. Then the wall of W given by $\gamma = 0$ is said to be of type α and is represented by the pair (W, α). For every closed path $W_0, W_1, \ldots, W_{p+1} = W_0$ with W_i adjacent to W_{i+1} there is a relation among the variables $(z(W, \alpha))$. If W_i and W_{i+1} are separated by a wall of type α_i then the relation is given by

$$\exp(z_p X_p) \exp(z_{p-1} X_{p-1}) \ldots \exp(z_0 X_0) = 1$$

where $z_i = z(W_i, \alpha_i)$ and $X_i = X_{-\alpha_i}$.

S' is defined to be the subvariety of S such that on each coordinate patch $S(B_\infty, B_0)$ and for each simple root α there is at least one chamber $W(\omega)$ for which $z(W(\omega), \alpha) \neq 0$. T_0 acts on $S'(B_\infty, B_0)$ and on affine r-space where r is the semisimple rank of G. The actions are given by

$$t : z(W, \alpha) \rightarrow \alpha(t) z(W, \alpha)$$

and

$$t : z(\alpha) \rightarrow \alpha(t^{-1}) z(\alpha).$$

The patches $S_1(B_\infty, B_0) = S'(B_\infty, B_0) \times_{T_0} \mathbb{A}^r$ piece together to form a variety S_1. There is a morphism from S_1 to S given locally by

$$(z(W, \alpha)), (z(\alpha)) \rightarrow (z(\alpha) z(W, \alpha))$$

$(z(W, \alpha)) \in S'(B_\infty, B_0), (z(\alpha)) \in \mathbb{A}^r$. We add subscripts $z_1(W, \alpha)$ to the variables in $S_1(B_\infty, B_0)$, $\{z_1(W, \alpha), z(\alpha)\}$ to distinguish them from their image $z_1(W, \alpha) z(\alpha)$ in S.

Now we describe the F-structure on the varieties. In chapters I through V we work with the variety over the algebraic closure, but beginning in chapter VI the F-structure will play an important role. We twist the ordinary action of Gal$(\bar{F}/F)$ on V^n. If we define an action of Gal$(\bar{F}/F)$ on Weyl chambers by $\sigma(\mathbb{B}(W)) = \mathbb{B}(\sigma(W))$ then the action of Gal$(\bar{F}/F)$ on V^n is given by $\sigma((B(W))) = (\sigma(B(\sigma^{-1}W)))$. The usual F-structure on G together with this twisted F-structure on V^n gives an F-structure on subvarieties of $G \times V^n$. There is a unique F-structure on S_1, X_1, etc. compatible with the F-structure just given to the subvarieties of $G \times V^n$. This action has been defined in such a way that the morphisms $T^0 \times T \backslash G \rightarrow X^0$ and $X \rightarrow T$ are defined over F.

Let M equal the Springer-Grothendieck variety $B \times_B G = \{(g, B) \in G \times V : g \in B\}$. There is a morphism $X \rightarrow M$ given by $(g, (B(W))) \rightarrow (g, B(W_+))$. The differential form $\prod(1 - \alpha^{-1}(\gamma)) \omega_T \wedge \omega_{T \backslash G}$ on $T \times T \backslash G$ when pulled back to M gives a G-invariant non-vanishing form ω_M on M. On the patch $M(B_\infty)$ of elements (g, B) such that B is opposite the Borel subgroup B_∞, we have $B = B_0^\nu, g =$

$(tn)^\nu$ with $t \in T_0, n \in N_0, \nu \in N_\infty$. The coefficients $t_1, \ldots, t_\ell;\ x_1, \ldots, x_p;\ \nu_1, \ldots, \nu_p$ of t, n and ν serve as coordinates on $M(B_\infty)$. The assumption that ω_M is G-invariant and non-vanishing forces ω_M to have the form

$$t_1^{a_1} \ldots t_\ell^{a_\ell} dt_1 \ldots dt_\ell dx_1 \ldots dx_p d\nu_1 \ldots d\nu_p.$$

Since we are only interested in the form near the identity, we may assume $|t_i| = 1$ and take ω_M to be

$$dt_1 \ldots dt_\ell dx_1 \ldots dx_p d\nu_1 \ldots d\nu_p$$

Consequently, we may take the forms ω_X and ω_Y on X^0 and Y^0 to be given by

$$dt_1 \ldots dt_\ell dx_1 \ldots dx_p d\nu_1 \ldots d\nu_p$$

and

$$d\lambda dx_1 \ldots dx_p d\nu_1 \ldots d\nu_p$$

respectively. The form ω_X is not defined over F in general but this is not a problem because there is always a constant $c \in \bar{F}^\times$ such that $c\omega_X$ is defined over F.

3. The Variety S^0.

It was mentioned in section 2, that there is a relation among the variables $z(W, \alpha)$ for every closed path $W_+ = W_0, \ldots, W_p, W_{p+1} = W_0$ selected through the Weyl chambers. These relations are not all independent. In fact, this section shows that all the relations are consequences of the relations that arise for rank two groups. This result is closely related to the fact that the Weyl group is a Coxeter group. Chapter III studies the rank two situation carefully. Building on lemma 3.1 and the results of Chapter III, Chapter IV will draw some general conclusions about the vanishing of principal values on divisors.

The rank two root systems occurring at the codimension two intersections of walls will be called *nodes*.

LEMMA 3.1. *Every relation among the coordinates* $(z(W, \alpha))$ *on a patch* $S^0(B_\infty, B_0)$ *of the variety of regular stars is a consequence of*

i) $z(W, \alpha) + z(W', \alpha) = 0$ *where* W *and* W' *are adjacent walls separated by a wall of type* α, *and*

ii) $\exp(z_p X_{-\alpha_p}) \ldots \exp(z_1 X_{-\alpha_1}) = 1$ *where* $W_1, W_2, \ldots, W_p$ *is the path around a node (so that* $p = 4, 6, 8, 12$ *according as the node is of type* $A_1 \times A_1$, A_2, B_2, G_2*) and* $z_1, \ldots, z_p$ *are the corresponding wall variables.*

PROOF 1. Consider any closed path $W_1, \ldots, W_{q+1} = W_1$. The chambers are separated by walls

$$(W_1, \alpha_1) = (W_2, \alpha_1), \ldots, (W_q, \alpha_q) = (W_1, \alpha_q).$$

Reflection in these walls corresponds respectively to elements $\omega_1, \ldots, \omega_q$ of the Weyl group, and $\omega_i W_i = W_{i+1}$ or

$$\omega_q \ldots \omega_1 W_1 = W_1.$$

The Weyl group acts simply transitively on the chambers so that $\omega_q \ldots \omega_1 = 1$. Any relation in the Weyl group is a consequence of the relations

i') $\omega_\alpha^2 = 1$

ii') $(\omega_\alpha \omega_\beta)^{m_{\alpha\beta}} = 1$ where $\pi/m_{\alpha\beta}$ is the angle formed by the walls (W_+, α) and (W_+, β) of the fundamental chamber.

We have the products

1) $\exp(z_q X_{-q}) \ldots \exp(z_1 X_{-1})$ $(X_{-i} =^{def} X_{-\alpha_i})$ and

2) $\omega_q \ldots \omega_1$.

Every time a relation (i') or (ii') is applied to (2) a similar relation (i) or (ii) can be applied to (1) to keep the length of both expressions the same. Repeated applications of (i') and (ii') will reduce the product in (2) to the identity, the identical process must then reduce the product in (1) to the identity.

PROOF 2. The Weyl chambers lie in $P = \mathbb{R}^n$. Let P^0 be the points of P lying in at most two walls. Then $P - P^0$ has codimension three so that P^0 is simply connected and has no non-trivial connected covering spaces. Construct a covering space as follows. Let each point of the covering space be given by a triple (x, W, p) where $x \in W$ and p is a path from W_+ to W. We now identify points. A necessary condition for (x, W, p) and (x', W', p') to be identified is that $x = x'$. If $x = x' \in P^0$, it must be true that W and W' are two chambers at a node (by the definition of P^0). Let $W = W_0, \ldots, W_q = W'$ be a path p'' joining W and W' such that for each i, W_i is a chamber at the node as well. Then (x, W, p) is to be identified with (x', W', p') provided that $x = x'$ and the path $p'^{-1}p''p$ from W_+ to W_+ (with composition of paths in the obvious sense) gives a relation that is a consequence of (i) and (ii). By conditions (i) and (ii) this condition is independent of the path p'' selected and is a local isomorphism. This covering space must be trivial and connected so every closed path gives a relation that is a consequence of (i) and (ii).

4. The Morphism $S_1 \to S$.

In this section we prove a proposition that will be used frequently and often implicitly in all that follows. The proposition was proved for A_2 and used in an essential way in [15]. It is the result required to insure that functions of compact support on G pull back to functions of compact support on the variety X_1. This proposition will be used in combinatorial arguments in chapters III and IV.

The following result will be used in the proof of the proposition and is stated here for reference. Here K is a field, R is a valuation ring with quotient field K, and $i : Spec(K) \to Spec(R)$ is the morphism induced by the inclusion $R \subseteq K$.

THEOREM 4.1. *(Valuative Criterion of Properness). Let $f : X \to Y$ be a morphism of finite type, with X noetherian. Then f is proper if and only if for every valuation ring R and for every morphism $Spec(K)$ to X and $Spec(R)$ to Y forming a commutative diagram*

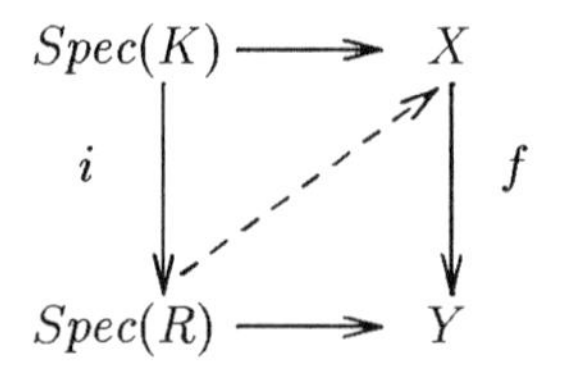

there exists a unique morphism $Spec(R) \to X$ making the whole diagram commutative.

PROOF. For details and a proof see [7].

PROPOSITION 4.2. *The morphism $p : S_1 \to S$ is proper and hence surjective.*

PROOF. We apply the valuative criterion of properness. Let η_1 be the image in S_1 of the unique point in $Spec(K)$. We select an affine patch on S_1 that intersects η_1 non-trivially. We may assume that the patch is given by a pair of opposite Borel subgroups (B_0, B_∞) and the conditions $z_1(W_\alpha, \alpha) \neq 0$ where we are given a Weyl chamber W_α for each simple root α, and $(z_1(W, \alpha)), (z(\alpha))$ are given representatives in $S'(B_\infty, B_0) \times \mathbb{A}^r$ for $S'(B_\infty, B_0) \times_{T_0} \mathbb{A}^r$. The condition $z_1(W_\alpha, \alpha) \neq 0$ is independent of the choice of representatives in $S'(B_\infty, B_0) \times \mathbb{A}^r$. On this affine patch the coordinate ring is generated by $z_1(W, \alpha)/z_1(W_\alpha, \alpha)$ $\forall$ (W, α) and $z(W_\alpha, \alpha) = z(\alpha) z_1(W_\alpha, \alpha)$ $\forall$ α. For each α let $W = W_\alpha^0$ be a choice of chamber for which $v(\varphi^*(z_1(W, \alpha)/z_1(W_\alpha, \alpha)))$ attains its minimum as W varies over chambers such that $\varphi^*(z_1(W, \alpha)/z_1(W_\alpha, \alpha)) \neq 0$. Here v is the valuation and φ is the morphism

$$\varphi : Spec(K) \to S_1.$$

Now η_1 intersects the affine patch $S_a \subseteq S_1(B_\infty, B_0)$ whose coordinate ring is generated by

$$z_1(W, \alpha)/z_1(W_\alpha^0, \alpha), z(W_\alpha^0, \alpha).$$

Then

$$v(\varphi^*(z_1(W, \alpha)/z_1(W_\alpha^0, \alpha))) \geq 0$$

for all (W, α). The map $S_1 \to S$ is given locally by

$$\begin{aligned}(z_1(W, \alpha)/z_1(W_\alpha^0, \alpha)), (z(W_\alpha^0, \alpha)) \to & (z_1(W, \alpha) z(W_\alpha^0, \alpha)/z_1(W_\alpha^0, \alpha)) \\ = & (z(W, \alpha)).\end{aligned}$$

Then the assumption of a morphism $Spec(R) \to S$ gives $v(\varphi^*(z(W, \alpha))) \geq 0$. In particular,

$$v(\varphi^*(z(W_\alpha^0, \alpha)) \geq 0.$$

Thus the image of the coordinate ring of S_a lies within the coordinate ring of R. This gives a morphism from $Spec(R)$ to S_a and hence to S_1. The uniqueness of the morphism is clear.

5. Cocycles.

The result of this section is essentially lemma 5.2 of [15]. We reproduce it here in a form more convenient for our applications. For every regular star $(t^g, (B(W))^g)$ there is a cocycle $\sigma(g)g^{-1}$ of $\mathrm{Gal}(\bar{F}/F)$ with values in $T(\bar{F})$. There is a character κ on $H^1(\mathrm{Gal}(\bar{F}/F), T) = H^1(T)$ used to determine the endoscopic group H such that the integrand f_1 on Y^0 is given by $\pi_1^*(f)m_\kappa(e)$ where $\pi_1^*(f)$ is the pullback of a locally constant function of compact support on G and $m_\kappa(e) = \kappa(\sigma(g)g^{-1})$.

PROPOSITION 5.2. *Let R be the field of rational functions of the variety of stars S. Then $m_\kappa(e)$ has an expression on a Zariski open set of the variety of regular stars S^0*

$$m_\kappa(e) = \kappa(t_\sigma(e)), \qquad t_\sigma(e) \in H^1(T), \qquad e \in S^0$$

where for each $\sigma \in \mathrm{Gal}(\bar{F}/F), t_\sigma$ belongs to $T(R)$.

PROOF. In the course of the proof we develop an expression from which t_σ may be calculated. We begin with a quasi-split group G_{qs} and an inner form G_{in}. We select a maximally split Cartan subgroup T_{qs} and a Borel subgroup B_{qs} both over F in G_{qs} with $T_{qs} \subseteq B_{qs}$. We select a Cartan subgroup T_{in} over F in G_{in}. Fix an isomorphism of G_{qs} with G_{in} over $\bar{F}$ which carries T_{qs} to T_{in}. We identify G_{qs} and G_{in} through this isomorphism. The two forms are distinguished by the actions of $\mathrm{Gal}(\bar{F}/F)$ on the groups. For $\sigma \in \mathrm{Gal}(\bar{F}/F)$ we write σ_{qs} and σ_{in} for the corresponding actions on the quasi-split and inner forms. For any Cartan subgroup T over F in G_{in}, select an element $h \in G(\bar{F})$ such that $T^h = T_{qs}$. Write $\sigma_{in}(h^{-1}) = w_\sigma h^{-1}$ with $w_\sigma \in N_G(T_{qs})$. Let $g \in T\backslash G_{in}(F)$. For g in a Zariski open set of G we can write $T^g = T_{qs}^{h^{-1}g} \subseteq B_{qs}^{h^{-1}g} = B_{qs}^{\nu}$ or $h^{-1}g = tn\nu$ with $t \in T_{qs}, n \in N_{qs}, \nu \in N_{qs\infty}$. N_{qs} is the unipotent radical of B_{qs} and $N_{qs\infty}$ is the unipotent radical of the Borel subgroup opposite to B_{qs} through T_{qs}. Thus $g = htn\nu = (hth^{-1})hn\nu = t'hn\nu$ with $t' \in T(\bar{F})$. Now

$$\sigma_{in}(g)g^{-1} = \sigma_{in}(t')\sigma_{in}(h)\sigma_{in}(n)\sigma_{in}(\nu)\nu^{-1}n^{-1}h^{-1}t'^{-1}$$

is a cocycle in $Z^1(T)$ which has the same class as

$$\sigma_{in}(h)\sigma_{in}(n)\sigma_{in}(\nu)\nu^{-1}n^{-1}h^{-1}.$$

We define a twisted action σ_* on T_{qs} by

$$\sigma_*(t) = \sigma_{in}(t)^{w_\sigma}, t \in T_{qs}.$$

Then if t_σ is a cocycle in $Z^1(T)$, we have

$$\tau_*(t_\sigma^h)t_\tau^h = \tau_{in}(t_\sigma)^{\tau_{in}(h)w_\tau}t_\tau^h = (\tau_{in}(t_\sigma)t_\tau)^h = (t_{\tau\sigma})^h.$$

Thus there is an identification of cocycles in T and twisted cocycles in T_{qs}.

$$h^{-1}\sigma_{in}(h)\sigma_{in}(n)\sigma_{in}(\nu)\nu^{-1}n^{-1} =$$

$$w_\sigma^{-1}\sigma_{in}(n)\sigma_{in}(\nu)\nu^{-1}n^{-1}$$

is then a cocycle in T_{qs} with this twisted action. Call this cocycle T_σ.

$$T_\sigma^{w_\sigma^{-1}} = \sigma_{in}(n)\sigma_{in}(\nu)\nu^{-1}n^{-1}w_\sigma^{-1}.$$

Since G_{qs} and G_{in} are inner forms, we may write

$$\sigma_{in}(g) = \mathbf{ad}\ A_\sigma^{-1}(\sigma_{qs}(g)),$$

where $\sigma \rightarrow A_\sigma$ is a cocycle with values in $N_{G_{qs}}(T_{qs})_{adj}$, the image of the normalizer in the adjoint group, with respect to the action σ_{qs}. Now

$$T_\sigma^{w_\sigma^{-1}} = A_\sigma^{-1}\sigma_{qs}(n)\sigma_{qs}(\nu)A_\sigma\nu^{-1}n^{-1}w_\sigma^{-1}$$

$$T_\sigma^{w_\sigma^{-1}A_\sigma^{-1}} = \sigma_{qs}(n)\sigma_{qs}(\nu)A_\sigma\nu^{-1}n^{-1}w_\sigma^{-1}A_\sigma^{-1}$$

EQUATION 5.3. $A_\sigma\nu^{-1}n^{-1}w_\sigma^{-1}A_\sigma^{-1} \in N_{\infty qs}N_{qs}T_\sigma^{w_\sigma^{-1}A_\sigma^{-1}}$.

These last two equations are the fundamental relation from which the function $m_\kappa(e)$ can be deduced for all reductive groups. The equation 5.3 determines T_σ as a rational function of the coefficients of

$$A_\sigma\nu^{-1}n^{-1}w_\sigma^{-1}A_\sigma^{-1}.$$

The cocycle w_σ measures the extent to which T_{in} and T are not isomorphic over F, and A_σ^{-1} measures the extent to which T_{in} and T_{qs} are not isomorphic over F. We give a few applications of this formula that will be useful in chapter VII when we carry out the transfer of the subregular germ for certain groups.

COROLLARY 5.4. *Suppose that $A_\sigma = 1$, $B_0 = B_{qs}$, $T_0 = T_{qs}$, and that w_σ is a simple reflection corresponding to the root α. Then $t_0T_\sigma = (z(W_+,\alpha))^{\alpha^\vee}$ for some element $t_0 \in T_{qs}(\bar{F})$ independent of the regular star.*

PROOF. Equation 5.3 gives $n^{-1}w_\sigma^{-1} \in N_{qs\infty}N_{qs}T_\sigma^{w_\sigma^{-1}}$. w_σ^{-1} differs from σ_α by an element of T_{qs} independent of the star. Here σ_α is the image in G of the reflection $\begin{pmatrix} 0 & 1 \\ -1 & 0 \end{pmatrix}$ in G_α where G_α is the rank one subgroup corresponding to the root α. We combine this element with T_σ and write $n^{-1}\sigma_\alpha \in N_{qs\infty}N_{qs}(t_0T_\sigma)^{\sigma_\alpha}$. Now by the 2 by 2 matrix calculation

$$\begin{pmatrix} 1 & -n_\alpha \\ 0 & 1 \end{pmatrix}\begin{pmatrix} 0 & 1 \\ -1 & 0 \end{pmatrix} = \begin{pmatrix} n_\alpha & 1 \\ -1 & 0 \end{pmatrix} = \begin{pmatrix} 1 & 0 \\ -1/n_\alpha & 1 \end{pmatrix}\begin{pmatrix} n_\alpha & 1 \\ 0 & 1/n_\alpha \end{pmatrix} =$$

$$\begin{pmatrix} 1 & 0 \\ -1/n_\alpha & 1 \end{pmatrix}\begin{pmatrix} 1 & n_\alpha \\ 0 & 1 \end{pmatrix}\begin{pmatrix} n_\alpha & 0 \\ 0 & 1/n_\alpha \end{pmatrix}.$$

EQUATION 5.5.

$$\epsilon_\alpha(-n_\alpha)\sigma_\alpha = \epsilon_{-\alpha}(-1/n_\alpha)\epsilon_\alpha(n_\alpha)n_\alpha^{\alpha^v}$$

where $\epsilon_{\pm\alpha}(X) = \exp(xX_{\pm\alpha})$.

From this it follows that $n^{-1}\sigma_\alpha \in N_\infty N_0 n_\alpha^{\alpha^v}$. Also

$$B(W(\sigma_\alpha)) = B_0^{\exp(z(W_+,\alpha)X_{-\alpha})\nu}$$

and $B(W(\sigma_\alpha)) = B_0^{\sigma_\alpha h^{-1}g} = B_0^{\sigma_\alpha n\nu}$. From this it follows that

$$B_0\sigma_\alpha n = B_0\exp(z(W_+,\alpha)X_{-\alpha}),$$

and by (5.5) it follows that $n_\alpha = 1/z(W_+,\alpha)$. The result follows from the relations $n_\alpha = 1/z(W_+,\alpha)$ and $\sigma_\alpha(\alpha^v) = -\alpha^v$.

REMARK 5.6. When $A_\sigma = 1$, for the classical groups we can recover T_σ from the equation

$$n^{-1}w_\sigma^{-1} \in N_{qs\infty}N_{qs}T_\sigma^{w_\sigma^{-1}}$$

by computing the principal minors of both sides noting that the principal minors of any matrix in $N_{qs\infty}N_{qs}$ are equal to one provided we choose a representation such that B_{qs} is upper triangular and T_{qs} is diagonal.

COROLLARY 5.7. *Suppose that A_σ is a simple reflection corresponding to a root α, and $\nu = \epsilon_{-\alpha}(\xi)^\alpha\nu$ with $^\alpha\nu \in N_\alpha$. Then T_σ is determined by the condition*

$$\sigma_\alpha\epsilon_{-\alpha}(-\xi)n^{-1}w_\sigma^{-1}\sigma_\alpha^{-1} \in N_{qs\infty}N_{qs}(t_1T_\sigma)^{w_\sigma^{-1}\sigma_\alpha^{-1}}$$

where $t_1 \in T_{qs}$ is independent of the star.

PROOF. This follows immediately from (5.3) if we note that

$$A_\sigma\ {}^\alpha\nu^{-1}A_\sigma^{-1} \in N_{qs\infty}.$$

6. The Data.

In this section we prove that the Igusa data exists for any reductive group, and show how to associate a unipotent conjugacy class to each divisor. Let G be a connected reductive group. We have morphisms φ, π_1, and ξ.

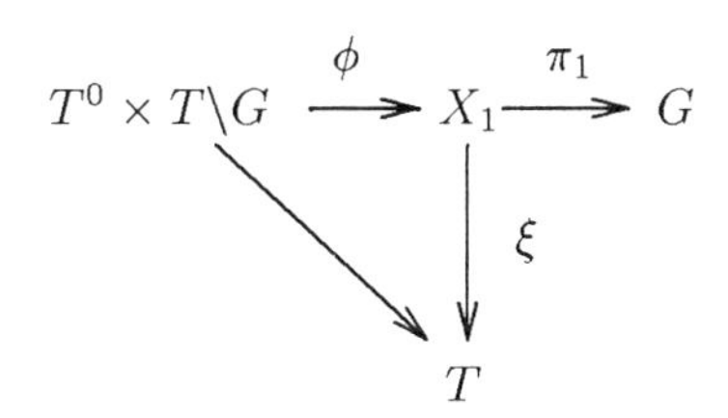

The maps are G-equivariant maps provided G acts on G by ad, on $T^0 \times T\backslash G$ by translation on the second factor, and trivially on T. The maps φ, ξ and π_1 are defined over F. $\xi \circ \varphi$ is projection on the first factor. So the diagram above commutes. $\pi_1(\varphi(t,g)) = \pi_1(t^g, (\mathbb{B}(W))^g) = t^g \in G$. The map π_1 is proper.

Now let Γ be a curve in T. The curve Γ is assumed to be a smooth curve which passes through the identity of T where its tangent is regular in the sense that it does not lie in a hyperplane defined by a root and which contains no other singular point. Let $\Gamma^0 = \Gamma\backslash\{0\}, Y_1^0 = \xi^{-1}(\Gamma^0)$, and let Y_1 be its closure in X_1. G then acts on Y_1 since ξ is a G-morphism. We also let 1 denote the element of $\Gamma \setminus \Gamma^0$. We have

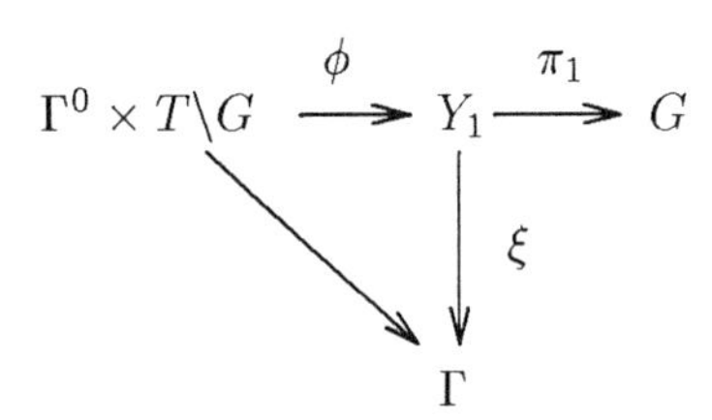

All morphisms are G-morphisms, and defined over F provided Γ is defined over F.

Replace Y_1 by a desingularization Y_Γ. The desingularization may be chosen to be G-equivariant, and the irreducible components of $\xi = 1$ may be assumed to have normal crossings [8]. Thus G acts on Y_Γ and all morphisms are G-morphisms.

$\xi^{-1}(1) \subseteq Y_\Gamma$ breaks up into a finite number of irreducible components. The expression *divisor* refers in this text to them. Conjugacy classes are taken to mean stable conjugacy classes unless indicated otherwise.

LEMMA 6.1. *Let E be a divisor in Y_Γ. There is a unique unipotent conjugacy class O in G such that $\pi_1(E)$ equals the closure of O in G.*

PROOF. Since G is connected, G fixes each divisor. $\pi_1(E)$ is also closed (by properness), irreducible, and G-invariant. So $\pi_1(E)$ is a union of conjugacy classes. If $\xi(x) = 1$ then $\pi_1(x)$ lies in a Borel subgroup B and $\pi_1(x)$ modulo N is 1. Thus $\pi_1(x)$ lies in N and is consequently unipotent. Call the unipotent classes in $\pi_1(E)$ $O_1, \ldots, O_j$ and their closures $X_1, \ldots, X_j$. $\pi_1(E) = X_i$ for some i by irreducibility. X_i determines the unipotent class uniquely for the classes in the closure are of strictly lower dimension.

If E is associated with the unipotent class O, call E an *O-divisor*. Similarly call E a *regular* or *subregular* divisor if the unipotent class associated to E is regular or subregular.

REMARK 6.2. Let $u, u' \in O$, then $u^g = u'$ for some g and $Y_\Gamma \to Y_\Gamma$ gives an isomorphism of the fibre $E(u)$ over u with the fibre $E(u')$ over u' whenever E is an O-divisor. This isomorphism is defined over F if u, u' lie in the same adjoint conjugacy class. By an adjoint conjugacy, we mean F-conjugacy in $G_{adj}(F)$. See [12] for details. By letting $u = u'$ we obtain an action of $C_G(u)$ on $E(u)$.

THEOREM 6.3. *Let G be reductive. There exists a variety Y_Γ proper over Y_1 such that the conditions for Igusa data are met.*

PROOF. By blowing up G-equivariantly if necessary we may assume the variety $\lambda = 0$ (i.e. $\xi = 1$) is the union of divisors with normal crossings. The only condition that has not been verified in [15] is that f locally has the form $\gamma\kappa_1(\mu_1)\dots\kappa_n(\mu_n)$ where $\mu_1, \dots, \mu_n$ are local F-coordinates on Y_Γ at a point p, and $\kappa_1, \dots, \kappa_n$ are characters of $F^\times$. The function γ must be locally constant at p and if κ_i is not the trivial character, $\mu_i = 0$ must define a divisor passing through p.

Fix a basis Λ for the characters of T. By (5.2), if $\chi \in \Lambda$ then $\chi(t_\sigma) \in R$ the field of rational functions on S. Working with a finite extension K of F and pulling $\chi(t_\sigma)$ back to the field of functions on Y_Γ we obtain finitely many prime divisors determined by the Weil divisors of $\chi(t_\sigma)$ for all $\chi \in \Lambda, \sigma \in \mathrm{Gal}(K/F)$. By blowing up if necessary, we may assume that these prime divisors together with the prime divisors determined by $\lambda = 0$ all have normal crossings. Let $\langle E \rangle$ be the set of prime divisors. Again blowing up if necessary, we may assume that a prime divisor of $\langle E \rangle$ has F-rational points if and only if it is defined over F. In the neighborhood on the p-adic manifold of an F-rational point p we may write

$$\chi(t_\sigma) = \alpha\mu_1^{a_1}\dots\mu_n^{a_n}$$

where $a_1, \dots, a_n, \alpha$ depend on χ and σ. We may assume that $\mu_1, \dots, \mu_n$ are local p-adic coordinates at p and that α is regular and invertible at p. It follows that t_σ has the form

$$t_\sigma : \sigma \to \mu_1^{\beta_{1\sigma}}\dots\mu_n^{\beta_{n\sigma}}\bar{t}_\sigma$$

for some cocharacters $\beta_{1\sigma}, \dots, \beta_{n\sigma}$ and $\bar{t}_\sigma$, where $\bar{t}_\sigma$ is regular at p. Since t_σ is a cocycle $\tau(t_\sigma)t_\tau = t_{\tau\sigma}$, that is

$$\mu_1^{\tau(\beta_{1\sigma})}\dots\mu_n^{\tau(\beta_{n\sigma})}\tau(\bar{t}_\sigma)\mu_1^{\beta_{1\tau}}\dots\mu_n^{\beta_{n\tau}}\bar{t}_\tau = \mu_1^{\beta_{1\tau\sigma}}\dots\mu_n^{\beta_{n\tau\sigma}}\bar{t}_{\tau\sigma}.$$

Rearranging:

$$\mu_1^{[\tau(\beta_{1\sigma})+\beta_{1\tau}-\beta_{1\tau\sigma}]}\dots\mu_n^{[\tau(\beta_{n\sigma})+\beta_{n\tau}-\beta_{n\tau\sigma}]} = \bar{t}_{\tau\sigma}\bar{t}_\tau^{-1}\tau(\bar{t}_\sigma^{-1}).$$

The right hand side is regular at p, so the left hand side must be as well. This forces

$$\tau(\beta_{i\sigma}) + \beta_{i\tau} = \beta_{i\tau\sigma}$$

for all i. Thus $\sigma \to \mu_i^{\beta_{i\sigma}}$ is a cocycle $\forall\ i$, and $\sigma \to \bar{t}_\sigma$ is as well. We then define the character κ_i on F^x by $\kappa_i(\mu) = \kappa(\mu^{\beta_{i\sigma}})$ where κ is the character on $H^1(T)$

defining the endoscopic group. This shows that the function f_1 has the correct form.

There is one last point to verify. It must be possible to choose the coordinates $\mu_1, \ldots, \mu_n$ in such a way that if $\mu_i = 0$ does not define a divisor then $\kappa_i = 1$. Since the divisors have normal crossings, we may assume that if E is a divisor then it is given locally by $\mu_i = 0$ for some i. But then if $\mu_j = 0$ does not define a divisor, points of $\mu_j = 0$, $\mu_i \neq 0$ $i \neq j$ are regular stars. The result then follows from the fact that $m_\kappa(e)$ is locally constant on the regular stars (cf. [15]).

II. COORDINATES AND COORDINATE RELATIONS

1. The Coordinates $x(W, \beta)$.

Before deriving any concrete results from the variety Y_Γ it will be necessary to develop coordinates charts on the variety. This section introduces coordinates $x(W,\beta)$ indexed by Weyl chambers W and positive roots β. They can be described as follows. Consider a point $p \in X_1 \subseteq G \times S_1$. Then locally $p \in B_0 \times S_1(B_\infty) \tilde{\rightarrow} B_0 \times S_1(B_\infty, B_0) \times N_\infty$. Write $p = (b, e, \nu)$. For each simple root α fix root vectors X_α and $X_{-\alpha}$ for $T_0 = B_0 \cap B_\infty$ in the Lie algebra of G such that $[X_\alpha, X_{-\alpha}] = H_\alpha$ with $\alpha(H_\alpha) = 2$. Fix an ordering on the positive roots then write $b \in B_0$ as

$$t \prod \exp(x_\beta X_\beta) \qquad \text{(ordered)}$$

with $t \in T_0$. To fix a convention, we agree that lower elements in the ordering appear to the left in the product. Then t and $x_\beta = x_\beta(b)$ are coordinates for b. If $e = (B_0^{n_w})$, then $b^{n_w^{-1}} \in B_0$ for all W. Define $x(W,\beta)$ to be $x_\beta(b^{n_w^{-1}})$. This definition depends on the order of the product. In concrete situations the order will always be specified. Notice, however, that $x(W,\alpha)$ for α simple is independent of the order.

2. The Coordinates $w(\beta)$.

This section defines a set of coordinates $w(\beta)$ on certain open patches $Y''(B_\infty)$ of the open set Y'' indexed by positive non-simple roots β. These coordinates will prove to be extremely useful on this open set. With them it will be possible to study the structure of those divisors in Y_Γ whose image in Y_1 meets Y''. The coordinates are easy to define; but it must be checked that they are truly regular coordinates on Y''. These verifications will be made in section 3.

Select the Borel subgroup B_∞ to be opposite to $\mathbb{B}(W_+)$. We work on the coordinate patch $Y^0(B_\infty)$. The restriction that B_∞ lie opposite $\mathbb{B}(W_+)$ is not a serious restriction. Although patches of this sort do not cover Y^0, translates of these patches by elements of G do cover Y^0 so that no structural information is lost by making the assumption that B_∞ is opposite $\mathbb{B}(W_+)$. We have maps:

$$T^0 \times T \backslash G \rightarrow X^0$$

$$(t, g) \rightarrow (t^g, (\mathbb{B}(W)^g)).$$

On $Y^0(B_\infty)$ we have $\mathbb{B}(W_+)^g = \mathbb{B}(W_+)^\nu$ for some $\nu \in N_\infty$, the unipotent radical of B_∞. Thus $g = t_0 n \nu$ for uniquely defined $t_0 \in T_0, n \in N_0$, and $\nu \in N_\infty$ where N_0 is the unipotent radical of $\mathbb{B}(W_+)$. Then on $Y^0(B_\infty, \mathbb{B}(W_+)) \times$

$N_\infty \rightrightarrows Y^0(B_\infty)$, $(t^g, (\mathbb{B}(W)^g))$ equals $(t^n, (\mathbb{B}(W)^n))^\nu$. Define $y(\beta)$ by $t^n = t\prod \exp(y(\beta)X_\beta)$. The definition of $y(\beta)$ depends on the order of the product. Suppose $\beta = \sum m(\alpha)\alpha$. Then define $w(\beta)$ by

$$w(\beta) = y(\beta)\prod z(\alpha)^{m(\alpha)}/\lambda$$

where $z(\alpha)$ is defined to be the quantity which makes $w(\alpha) = 1$ for α simple. That is, $z(\alpha) = \lambda/y(\alpha)$.

This gives the definition of $w(\beta)$ on an open set of $Y_1(B_\infty)$. This definition depends on the order the product defining $y(\beta)$. In applications the order must be specified.

It must be checked that $w(\beta)$ extends to a regular coordinate on $Y''(B_\infty)$. Formulas will be given relating the coordinates $w(\beta)$ to t, n, and the coordinates

$$z_1(W,\alpha) =^{def} z(W,\alpha)/z(\alpha).$$

These topics are treated in the next few sections.

3. The Extension of $w(\beta)$ *to* Y''.

This section shows that the coordinates $w(\beta)$ are regular on $Y''(B_\infty)$. Before proving the result I state a well known lemma that will be needed in the proof.

LEMMA 3.1. *Let α and β be positive roots, and let Ψ be the set of roots of the form $r\alpha + s\beta$ (r, s positive integers). Fix vectors X_γ. Then the commutator $(\exp(xX_\alpha), \exp(yX_\beta))$ equals $\prod \exp(c_{\alpha\beta\gamma}x^r y^s X_\gamma)$, where the product is taken over all $\gamma = r\alpha + s\beta \in \Psi$ (in some fixed order) and where $c_{\alpha\beta\gamma}$ are constants independent of x and y.*

PROOF. [10,32.5].

The following result is independent of the order selected on the roots to define $y(\beta)$.

LEMMA 3.2. *The coordinates $w(\beta)$ are regular on $Y''(B_\infty)$. The coordinates $w(\beta)$ may be expressed as a function of $\{t, (z(\alpha)), (z_1(W,\alpha))\}$. As such they are actually independent of the coordinates $\{z(\alpha)\}$.*

In the course of the proof we will prove a second lemma. Write for any element $n \in N_0$ and $t \in T$

$$n = \prod \exp(n_\beta X_\beta) \qquad \text{(ordered)}$$
$$t^n = t\prod \exp(y(\beta)X_\beta) \qquad \text{(ordered)}$$

Any order on the roots may be selected, but it must be the same for both products. Solve these equations for $y(\beta)$ in terms of the variables $\{n_\alpha, \alpha^{-1}(t)\}$ (α positive). We obtain an expression of the form:

$$y(\beta) = \sum c_{\beta_1\ldots\beta_n}(t) n_{\beta_1} n_{\beta_2}\ldots n_{\beta_n}$$

where the sum ranges over the set $\beta_1 + \beta_2 + \ldots + \beta_n = \beta$.

LEMMA 3.3. *$c_{\beta_1\ldots\beta_n}(t)$ is a sum of terms of the form $(1-\gamma_1^{-1}(t))\gamma_2(t)$ where γ_1 is a root and γ_2 is a linear combination of roots. Also $c_\beta(t)$ equals $(1-\beta^{-1}(t))$. In particular $y(\alpha)=(1-\alpha^{-1})n_\alpha$ for α simple.*

PROOF. The positive roots can be numbered according to the ordering on them: $\alpha_1,\ldots,\alpha_k$ so that

$$n=\prod \epsilon_i(n_i) \text{ with } \epsilon_i=\epsilon_{\alpha_i}, n_i=n_{\alpha_i}, \text{ and } \epsilon_\alpha(x)=\exp(xX_\alpha).$$

Then $t^{-1}n^{-1}tn$ is given by

$$\epsilon_k(-\alpha_k^{-1}(t)n_k)\ldots\epsilon_1(-\alpha_1^{-1}(t)n_1)\epsilon_1(n_1)\ldots\epsilon_k(n_k).$$

The innermost terms combine to give $\epsilon_1((1-\alpha_1^{-1}(t))n_1)$. This term can be pulled through the product to the left. By (3.1), doing so will only add terms whose dependence on T has the form $(1-\alpha_1^{-1}(t))\gamma(t)$ where γ lies in the coordinate ring of T. By repeatedly pulling out the innermost term of the product, we arrive at the result. It is clear from this procedure that $c_\beta(t)$ equals $(1-\beta^{-1}(t))$. This completes the proof of lemma 3.3.

We continue with the proof of (3.2). n_β is a function on $Y^0(B_\infty)\subseteq G\times S^0(B_\infty,B_0)\times N_\infty$. It actually depends only on the second factor so that n_β is a function on $S^0(B_\infty,B_0)$. By the inclusion

$$S^0(B_\infty,B_0)\subseteq S''(B_\infty,B_0),$$

$n_\beta\prod z(\alpha)^{m(\alpha)}$ is then a rational function on $S''(B_\infty,B_0)$.

LEMMA 3.4. *$n_\beta\prod z(\alpha)^{m(\alpha)}$ considered as a rational function on $S''(B_\infty,B_0)$ is regular and depends only on the coordinates*

$$z_1(W,\alpha)=z(W,\alpha)/z(\alpha)$$

and not on the coordinates $z(\alpha)$.

REMARK. This lemma will complete the proof of lemma 3.2, for

$$w(\beta)=y(\beta)(\prod z(\alpha)^{m(\alpha)})/\lambda=\sum[c_{\beta_1\ldots\beta_n}(t)/\lambda]\prod(n_{\beta_i}\prod z(\alpha)^{m_i(\alpha)})$$

where $\beta_i=\sum m_i(\alpha)\alpha$ and $c_{\beta_1\ldots\beta_n}(t)/\lambda$ is regular at $\lambda=0$.

PROOF OF 3.4. The matrices n_w depend only on $(z(W,\alpha))$. Since $z(W,\alpha)=z(\alpha)z_1(W,\alpha)$ they depend on $z_1(W,\alpha)$ and $z(\alpha)$. Recall that the matrix n_w in N_∞ is defined by the condition $\mathbb{B}(W)^{g\nu^{-1}}=B_0^{n_w}$. On our coordinate patch

$$B_0=\mathbb{B}(W_+)=\mathbb{B}$$

and $g = t_0 n\nu$. The condition

$$\mathbb{B}(W)^n = \mathbb{B}^{n_w} \ \forall \ W$$

allows one to express n_β in terms of the variables $\{z(\alpha), z_1(w,\alpha)\}$. The torus T_0 acts on the points $e = (\mathbb{B}(W)^{n\nu}) = (\mathbb{B}^{n_w \nu})$ by

$$e \rightarrow e^{t_0} = (\mathbb{B}(W)^{n\nu t_0}) = (\mathbb{B}^{n_w \nu t_0}),$$

$t_0 \in T_0(\bar{F})$. The coordinates of $e^{t_0} = (\mathbb{B}(W)^{n'\nu'}) = (\mathbb{B}^{n'_w \nu'})$ are clearly given by $\nu' = adt_0^{-1}(\nu)$, $n' = adt_0^{-1}(n)$, $n'_w = adt_0^{-1}(n_w)$, or

$$n'_\beta = \beta(t_0^{-1}) n_\beta, \quad z'(W,\alpha) = \alpha^{-1}(t_0^{-1}) z(W,\alpha) = \alpha(t_0) z(W,\alpha).$$

For any choices of $z(\alpha) \in \bar{F}^\times$, t_0 can be selected to give $\alpha(t_0^{-1}) = z(\alpha)$ for all α. Then $\beta(t_0^{-1}) = \prod z(\alpha)^{m(\alpha)}$ if $\beta = \sum m(\alpha)\alpha$. Write $\underline{n}_\beta$ for the rational function of the variables $(z(\alpha)), (z_1(W,\alpha))$ described above, n_β for the value of $\underline{n}_\beta$ at $(z(\alpha)), (z_1(W,\alpha))$ and n'_β for the value of $\underline{n}_\beta$ at $(1^r, (z'(W,\alpha))) = (1^r, (\alpha(t_0) z(W,\alpha))) = (1^r, (z(\alpha)^{-1} z(W,\alpha))) = (1^r, (z_1(W,\alpha)))$. ($1^r$ denotes the vector in $\mathbb{A}^r$ whose components are all equal to one.) This gives the needed independence:

$$\underline{n}_\beta(1^r, (z_1(W,\alpha))) = \prod z(\alpha)^{m(\alpha)} \underline{n}_\beta((z(\alpha)), (z_1(W,\alpha)))$$

for

$$\begin{aligned} \underline{n}_\beta(1^r, (z_1(W,\alpha)))/(\prod z(\alpha)^{m(\alpha)}) &= n'_\beta/(\prod z(\alpha)^{m(\alpha)}) \\ &= \beta(t_0) n'_\beta \\ &= n_\beta \\ &= \underline{n}_\beta((z(\alpha)), (z_1(W,\alpha))). \end{aligned}$$

Finally, we check that $n_\beta \prod z(\alpha)^{m(\alpha)}$ is regular on $S''(B_\infty, B_0)$. The point $((z_1(W,\alpha)), \nu) \in S^0(B_\infty, B_0) \times N_\infty$ describes a regular star e. Since it is a regular star there is a unique $n_0 \in N_0$ such that $e = (\mathbb{B}(W)^{n_0 \nu})$. It follows that the coefficients $n_{0\beta}$ of n_0 are regular functions of $(z_1(W,\alpha))$. But $n_{0\beta}(z_1(W,\alpha)) = \underline{n}_\beta(1^r, (z_1(W,\alpha))) = n_\beta \prod z(\alpha)^{m(\alpha)}$ so that $n_\beta \prod z(\alpha)^{m(\alpha)}$ too is regular. The proofs of the lemmas 3.2 and 3.4 are now complete.

4. The Coordinate Ring.

For $(g, (B(W)))$ in $Y''(B_\infty)$ write $(g, (B(W))) = (b, (B_0^{n_w}))^\nu$ where $b \in B_0 = \mathbb{B}(W_+)$. For the next proposition it is important to work on the affine patch $Y''(B_\infty)$. We let λ denote the pullback to Y_Γ (or any related variety) of a local parameter on Γ.

PROPOSITION 4.1.

a) *The subring of the coordinate ring of $Y''(B_\infty, \mathbb{B}(W_+)$ generated by λ and $\{z_1(W,\alpha) : \forall \ (W,\alpha)\}$ is contained in the subring generated by the coordinates $\{w(\gamma) : \gamma > 0, \gamma$ not simple$\}$ and λ.*

b) The coefficients of $b, \{z(\alpha) : \alpha$ *simple*$\}$, λ, *and* $\{w(\gamma) : \gamma$ *positive but not simple*$\}$ *are regular and together generate the coordinate ring of* $Y''(B_\infty, \mathbb{B}(W_+))$.

PROOF. Let R be the ring generated by λ and $\{w(\gamma)\}$. We have seen that λ and $\{w(\gamma)\}$ are regular.

We have

$$w(\beta) = \sum [c_{\beta_1 \ldots \beta_n}(t)/\lambda] \prod (n_{\beta_i} \prod z(\alpha)^{m_i(\alpha)})$$

$\beta = \sum m(\alpha)\alpha$. Define $\tilde{n}_\beta = n_\beta \prod z(\alpha)^{m(\alpha)}$ for $\beta = \sum m(\alpha)\alpha$. Then

$$w(\beta) = \sum [c_{\beta_1 \ldots \beta_n}(t)/\lambda] \prod \tilde{n}_{\beta_i} = [(1 - \beta^{-1}(t))/\lambda]\tilde{n}_\beta + \ldots$$

where the omitted terms all contain more than one $\tilde{n}_{\beta_i}$ as a factor. We show that $\tilde{n}_\beta$ lies in R. By induction we may assume that $\tilde{n}_\gamma$ lies in the ring R for $m_\gamma < m_\beta$ where $m_\gamma = \sum m(\alpha), \gamma = \sum m(\alpha)\alpha$. We have

$$w(\beta) = [(1 - \beta^{-1}(t))/\lambda]\tilde{n}_\beta + x \text{ with } x \in R$$

$$\text{and } \tilde{n}_\beta = (w(\beta) - x)(\lambda/(1 - \beta^{-1}(t))).$$

This belongs to R since the curve Γ is assumed to be regular at $\lambda = 0$ so that $\lambda/(1-\beta^{-1})$ is regular at $\lambda = 0$. Note that at the first step of the induction $x = 0$ and $\tilde{n}_\alpha = \lambda/(1 - \alpha^{-1}(t))$ for α simple.

Now let $\tilde{n} = \prod \epsilon_\beta(\tilde{n}_\beta)$. There is a regular star given by $(B_0^{\omega\tilde{n}})$. [15,3.1] gives an algorithm to solve for n_w and hence for $(z_1(W, \alpha))$ where $B_0^{n_w} = B_0^{\omega\tilde{n}}, W = W(\omega)$ provided $B_0^{\omega\tilde{n}}$ is opposite B_∞ for all ω. On the affine patch $Y''(B_\infty)$ this is true by definition. This proves (a).

(b) We first show that $z(\alpha)$ is regular.

$$z(\alpha) = \lambda/y(\alpha) = \lambda/((1 - \alpha^{-1})n_\alpha)$$

so that $z(\alpha)$ is regular provided $1/n_\alpha$ is regular. By the comment following (I.5.5), $1/n_\alpha = z(W_+, \alpha)$ which is certainly regular.

By (a) $z_1(W, \alpha)$ for all (W, α) lies in the ring generated by

$$\{w(\gamma) : \gamma\}, \{z(\alpha) : \alpha\}, \lambda.$$

But $\{z(\alpha)\}, \{z_1(W, \alpha)\}$, λ and the coefficients of b generate the coordinate ring of $Y''(B_\infty, B_0)$.

PROPOSITION 4.2. *Write* $b = t \cdot \prod \epsilon_\beta(x(\beta))$ *then on* $Y''(B_\infty, B_0)$ *the following equations hold:*

$$w(\alpha) = 1 : \alpha \textit{ simple}$$

$$\lambda w(\beta) = x(\beta) \prod z(\alpha)^{m(\alpha)} : \beta = \sum m(\alpha)\alpha$$

$$w(\gamma)x(\beta) = w(\beta)x(\gamma) \prod z(\alpha)^{m(\alpha)} : \gamma - \beta = \sum m(\alpha)\alpha.$$

PROOF. Referring to the definition of $w(\beta)$ we must have $x(\beta) = y(\beta)$ because $b = t^n$ on $Y^0(B_\infty, B_0)$.

5. A Computation of $t^{-1}n^{-1}tn$.

This section continues the discussion of the variables $w(\gamma)$. The purpose of the section is to derive formulas relating $w(\gamma)$ to t and n. That is, we compute the product $t^{-1}n^{-1}tn$. In lemma 3.3 it was shown that if $t^{-1}n^{-1}tn = \prod \exp(y(\beta)X_\beta)$ then each coefficient $y(\beta)$ is a sum of terms of the form

$$c_{\beta_1\ldots\beta_n}(t)n_{\beta_1}\ldots n_{\beta_n} \quad \text{where} \quad \beta_1 + \ldots + \beta_n = \beta.$$

What follows is a computation of the $c_{\beta_1\ldots\beta_n}s$.

All unipotent elements of a reductive group belong to the derived subgroup G_{der}, so $t^{-1}n^{-1}t$ and n can be taken in G_{der} to calculate the product. In fact, we can work in any cover, for $\prod \exp(y(\beta)X_\beta)$ is unchanged if t is changed by a central element.

We illustrate with the group A_n. Order the roots as follows. Let $\alpha_r + \ldots \alpha_{r+s}$ be associated with the pair $(n-r, s)$. Then order the roots by the lexicographical ordering on the ordered pairs. The smallest few roots for A_n will be $\alpha_n; \alpha_{n-1}, \alpha_{n-1}+\alpha_n; \alpha_{n-2}, \alpha_{n-2}+\alpha_{n-1}$, etc. The order is illustrated by the following diagram.

$$\begin{array}{cccc} \cdot & \cdot & \cdot & \cdot \\ 7 & 8 & 9 & 10 \\ & 4 & 5 & 6 \\ & & 2 & 3 \\ & & & 1 \end{array}$$

LEMMA 5.1. *With the ordering on the roots just given, $y(\beta)$ is a sum of the terms*

$$(-1)^j\beta^{-1}(t)(1-\beta_j(t))n_{\beta_1}\ldots n_{\beta_j}$$

where $\beta = \alpha_r + \ldots + \alpha_{r+s}, \beta_i = \alpha_{a_{i-1}-1} + \ldots + \alpha_{a_i} (a_{i-1} - 1 \le a_i)$ for $i = 1, \ldots, j$ and $r + 1 = a_0, a_j = r + s$.

PROOF. Send the exponential $\epsilon_\gamma(x) : \gamma = \alpha_r + \ldots + \alpha_{r+s}$ to the matrix $I + xe_{r,r+s+1} \in SL(n+1)$ where I is the identity and $e_{r,r+s+1}$ is the $n+1$ by $n+1$ matrix with 1 in the $(r, r+s+1)st$ position and 0 elsewhere. Note that $e_{ij}e_{\ell m} = \delta_{j\ell}e_{im}$. With the ordering selected $e_{ij}e_{\ell m} = 0$ provided e_{ij} corresponds to a root preceding that of $e_{\ell m}$ for then $\ell \le i < j$ so that $\delta_{j\ell} = 0$. Define a matrix $Y = (y_{ij})$ by $y_{r,r+s+1} = y(\gamma)$ $(r \ge 1, s \ge 0)$, $y_{ij} = 0$ $(i > j)$, and $y_{ii} = 1$ $(i = 1, \ldots, n)$. Then $\prod \epsilon_\beta(y(\beta))$ is sent to

$$\prod (I + y_{r,r+s+1}e_{r,r+s+1}) = I + \sum y_{r,r+s+1}e_{r,r+s+1} = (y_{ij}).$$

I claim that if $n = (n_{ij})$ then $m = (m_{ij})$, given by

$$m_{ij} = \begin{cases} \sum(-1)^{k+1}n_{ip_1}n_{p_1p_2}\cdots n_{p_kj} & \text{if } i < j \\ (\text{sum over all } i < p_1 < p_2 < \cdots p_k < j) & \\ 1 \quad \text{if } i = j & \\ 0 \quad \text{if } i > j & \end{cases}$$

is the inverse of n. For

$$\sum_j n_{ij}m_{jk} = \sum_{k\geq j\geq i} n_{ij}m_{jk} = \begin{cases} 0 & \text{if } i > k \\ 1 & \text{if } i = k \\ \sum_{j=i}^{k} n_{ij}m_{jk} & \text{if } i < k \end{cases}$$

Also

$$\begin{aligned} \sum_{j=i}^{k} n_{ij}m_{jk} &= m_{ik} + \sum_{j=i+1}^{k} \sum (-1)^{\ell+1} n_{ij}n_{jp_1}n_{p_1p_2}\cdots n_{p_\ell k} \\ &= m_{ik} + (-1)m_{ik} = 0. \end{aligned}$$

Notice too that

$$(t^{-1}mt)_{ij} = \begin{cases} \sum(-1)^{k+1}\beta'^{-1}(t)n_{ip_1}n_{p_1p_2}\cdots n_{p_kj} & \text{if } i < j \\ \text{where } \beta' = \alpha_i + \ldots + \alpha_{j-1} & \\ 1 \quad \text{if } i = j & \\ 0 \quad \text{if } i > j. & \end{cases}$$

Set $m'_{ij} = (t^{-1}mt)_{ij}$

$$(t^{-1}n^{-1}tn)_{ij} = \sum m'_{ik}n_{kj} = \begin{cases} 0 & \text{if } i > j \\ 1 & \text{if } i = j \\ \sum_{k=i}^{j} m'_{ik}n_{kj} & \text{if } i < j \end{cases}$$

If $i < j$,

$$\begin{aligned} \sum_{k=i}^{j} m'_{ik}n_{kj} &= m'_{ij} + \sum_{k=i}^{j-1}\sum(-1)^{\ell+1}\beta'^{-1}(t)n_{ip_1}n_{p_1p_2}\cdots n_{p_\ell k}n_{kj} \\ &\qquad (\beta' = \alpha_i + \ldots + \alpha_{k-1}) \\ &= \sum(-1)^{m+1}\beta^{-1}(t)n_{iq_1}n_{q_1q_2}\cdots n_{q_mj} + \\ &\qquad \sum(-1)^{\ell+1}\beta'^{-1}(t)n_{ip_1}n_{p_1p_2}\cdots n_{p_{\ell+1}j} \\ &\qquad (\beta = \alpha_i + \cdots + \alpha_{j-1}, \beta' = \alpha_i + \cdots + \alpha_{p_{\ell+1}-1}) \\ &= \sum(-1)^{m+1}(\beta^{-1}(t) - \beta'^{-1}(t))n_{iq_1}n_{q_1q_2}\cdots n_{q_mj} \\ &\qquad (\beta = \alpha_i + \cdots + \alpha_{j-1}, \beta' = \alpha_i + \cdots + \alpha_{q_m-1}) \\ &\qquad (i < q_1 < \cdots < q_m < j) \\ &= \sum(-1)^m\beta^{-1}(t)(1 - \beta_m(t))n_{\beta_1}n_{\beta_2}\cdots n_{\beta_m}. \end{aligned}$$

6. A Technical Lemma.

This section indicates how to calculate the coefficients n_γ as functions of the variables $\{z(W,\alpha)\}$. The method presented here has the advantage of working for all reductive groups. In practice, however, it is laborious. The coefficients n_γ for the classical groups can be calculated directly without using the method presented here. Section 7 will then compute the coefficients n_γ for the group G_2 using this method. Section 8 will carry out the computation for $SL(n)$ using an easier method. The computation of n_γ can be combined with the expressions for $w(\gamma)$ in terms of t and n to give expressions for $w(\gamma)$ in terms of the variables $\{z_1(W,\alpha)\}$ and t.

We need a more precise formulation of [15,5.3]. Let $\omega \in \Omega$ be the Weyl group element such that $W = \omega^{-1}W_+ = W(\omega)$. Let σ_ω be a representative of ω in the normalizer of T_0. If $\omega = \sigma_\alpha$ a simple reflection we let σ_α also denote the representative

$$\begin{pmatrix} 0 & 1 \\ -1 & 0 \end{pmatrix}$$

in G through G_α where G_α is the rank one subgroup of G associated with the root α of T_0 [10]. We have $\mathbb{B}(W)^n = \mathbb{B}^{\omega n} = \mathbb{B}^{n_w}$, so that on $Y^0(B_\infty, \mathbb{B}(W_+))$ $tn_0\sigma_\omega n = n_w$ for some $t \in T_0$ and $n_0 \in N_0$ the unipotent radical of $\mathbb{B}$. For every root γ fix a vector X_γ in the root space of γ. Implicit in [15,5.3] is a formula for t. We need to compute n_0 as well. Let $tv\sigma_\omega m = n_w$ and $t'v'\sigma_{\omega'}m' = n_{w'}$ where

$$\sigma_\alpha\omega' = \omega; n_w = \exp(zX_{-\alpha})n_{w'}$$

$$W = W(\omega), W' = W(\omega'), \omega'\gamma = \alpha \quad (\gamma \text{ positive})$$

$$v, v' \in N_0$$

$$m' \text{ lies in } N_{\omega'}, \text{ and } m \text{ lies in } N_\omega.$$

We assume that σ_ω and $\sigma_{\omega'}$ are chosen so that $\sigma_\alpha\sigma_{\omega'} = \sigma_\omega$. Here N_ω is the connected subgroup of N_0, the unipotent radical of B_0, whose Lie algebra is spanned by

$$\{X_\alpha | \alpha > 0, \omega \cdot \alpha < 0\}.$$

Furthermore write $v' = \exp(y_\alpha X_\alpha)^\alpha v'$ where ${}^\alpha v' \in N_\alpha$ the unipotent radical of the parabolic subgroup $\mathbb{P}_\alpha$ associated to the simple root α; and define u by $\sigma_{\omega'}X_\gamma\sigma_{\omega'}^{-1} = uX_\alpha$.

LEMMA 6.1. *With notation as above, the element n_w has a decomposition of the form $n_w = tv\sigma_\omega m$. There is a unique element m_γ such that $m = \exp(m_\gamma X_\gamma)m'$. Furthermore,*

$$\alpha(t')z(y_\alpha - um_\gamma) + 1 = 0$$

$$t = t'(-\alpha(t')z)^{-\alpha^v}, \quad \text{and}$$

$$v = \exp(\alpha(t')zX_\alpha)\sigma_\alpha \exp(um_\gamma X_\alpha)^\alpha v' \exp(-um_\gamma X_\alpha)\sigma_\alpha^{-1}.$$

REMARK. The result is highly dependent on the order of the products, on the choices of root vectors, and the choices of Weyl group representatives.

PROOF. The existence of the decomposition of n_w will follow from the calculations giving formulas for t and v.

$$n_w = tv\sigma_\omega m = tv\sigma_\alpha\sigma_{\omega'}\exp(m_\gamma X_\gamma)m',$$
$$n_w = \exp(zX_{-\alpha})n_{w'} = \exp(zX_{-\alpha})t'v'\sigma_{\omega'}m'$$

So

$$tv\sigma_\alpha\sigma_{\omega'}\exp(m_\gamma X_\gamma)\sigma_{\omega'}^{-1} = \exp(zX_{-\alpha})t'v' = t'\exp(\alpha(t')zX_{-\alpha})v'$$

$$\begin{aligned} tv &= t'\exp(\alpha(t')zX_{-\alpha})\exp(y_\alpha X_\alpha)^\alpha v'\exp(-m_\gamma uX_\alpha)\sigma_\alpha^{-1} \\ &= t'\exp(\alpha(t')zX_{-\alpha})\exp((y_\alpha - um_\gamma)X_\alpha)\sigma_\alpha^{-1}v'' \end{aligned}$$

where

$$v'' \overset{(def)}{=} \sigma_\alpha\exp(um_\gamma X_\alpha)^\alpha v'\exp(-um_\gamma X_\alpha)\sigma_\alpha^{-1} \in N_\alpha.$$

Now

$$\begin{pmatrix} 1 & 0 \\ A & 1 \end{pmatrix}\begin{pmatrix} 1 & B \\ 0 & 1 \end{pmatrix}\begin{pmatrix} 0 & -1 \\ 1 & 0 \end{pmatrix} = \begin{pmatrix} * & * \\ 0 & * \end{pmatrix}$$

forces $AB + 1 = 0$. So

$$\alpha(t')z(y_\alpha - um_\gamma) + 1 = 0.$$

Also

$$\begin{pmatrix} 1 & 0 \\ A & 1 \end{pmatrix}\begin{pmatrix} 1 & -1/A \\ 0 & 1 \end{pmatrix}\begin{pmatrix} 0 & -1 \\ 1 & 0 \end{pmatrix} = \begin{pmatrix} 1 & -1/A \\ A & 0 \end{pmatrix}\begin{pmatrix} 0 & -1 \\ 1 & 0 \end{pmatrix}$$

$$= \begin{pmatrix} -1/A & -1 \\ 0 & -A \end{pmatrix} = \begin{pmatrix} -1/A & 0 \\ 0 & -A \end{pmatrix}\begin{pmatrix} 1 & A \\ 0 & 1 \end{pmatrix}$$

So

$$\begin{aligned} tv &= t'(-\alpha(t')z)^{-\alpha^\vee}\exp(\alpha(t')zX_{-\alpha})v'' \\ t &= t'(-\alpha(t')z)^{-\alpha^\vee} \\ v &= \exp(\alpha(t')zX_\alpha)v''. \end{aligned}$$

7. Application to G_2.

As an application of the lemma we compute the coefficients n_γ for the group G_2. These coefficients will be needed for calculations carried out in chapter III. Let α be short, β long so that the positive roots are $\alpha, \beta, \alpha+\beta, 2\alpha+\beta, 3\alpha+\beta, 3\alpha+2\beta$. Products will be taken in this order. The order in which the roots are taken here does not correspond to the ordering forced upon the roots in lemma 6.1; so we must convert from one ordering to another. To facilitate the computations to be carried out in this section, we list the action of the reflections σ_α and σ_β on the roots. We also list the sets R_ω, where R_ω is defined to be $\{\beta > 0 | \omega\beta < 0\}$.

$\sigma_\beta x$	x	$\sigma_\alpha x$
$\alpha+\beta$	α	$-\alpha$
$-\beta$	β	$3\alpha+\beta$
α	$\alpha+\beta$	$2\alpha+\beta$
$2\alpha+\beta$	$2\alpha+\beta$	$\alpha+\beta$
$3\alpha+2\beta$	$3\alpha+\beta$	β
$3\alpha+\beta$	$3\alpha+2\beta$	$3\alpha+2\beta$

ω	R_ω
σ_α	$\{\alpha\}$
$\sigma_\beta\sigma_\alpha$	$\{\alpha, 3\alpha+\beta\}$
$\sigma_\alpha\sigma_\beta\sigma_\alpha$	$\{\alpha, 3\alpha+\beta, 2\alpha+\beta\}$
σ_β	$\{\beta\}$
$\sigma_\alpha\sigma_\beta$	$\{\beta, \alpha+\beta\}$
$\sigma_\beta\sigma_\alpha\sigma_\beta$	$\{\beta, \alpha+\beta, 3\alpha+2\beta\}$

Note also that $\alpha+\beta = \sigma_\beta\alpha, 2\alpha+\beta = \sigma_\alpha\sigma_\beta\alpha, 3\alpha+\beta = \sigma_\alpha\beta, 3\alpha+2\beta = \sigma_\beta\sigma_\alpha\beta$.

To continue our preparations, we also list some products in G_2 that will be needed. Let $\gamma = \alpha+\beta, \delta = 2\alpha+\beta, \epsilon = 3\alpha+\beta, \zeta = 3\alpha+2\beta$. Define the vectors $X_\gamma, X_\delta, X_\epsilon$, and X_ζ by $X_\gamma = \text{Ad } \sigma_\beta(X_\alpha), X_\delta = \text{Ad } \sigma_\alpha(X_\gamma), X_\epsilon =$ - Ad $\sigma_\alpha(X_\beta), X_\zeta = \text{Ad } \sigma_\beta(X_\epsilon)$. Define the structure constants $a, b, \ldots, j$ by the relations

$$\epsilon_\beta(y)\epsilon_\alpha(x) = \epsilon_\alpha(x)\epsilon_\beta(y)\epsilon_\gamma(axy)\epsilon_\delta(bx^2y)\epsilon_\epsilon(cx^3y)\epsilon_\zeta(dx^3y^2)$$
$$\epsilon_\gamma(y)\epsilon_\alpha(x) = \epsilon_\alpha(x)\epsilon_\gamma(y)\epsilon_\delta(exy)\epsilon_\epsilon(fx^2y)\epsilon_\zeta(gxy^2)$$
$$\epsilon_\delta(y)\epsilon_\alpha(x) = \epsilon_\alpha(x)\epsilon_\delta(y)\epsilon_\epsilon(hxy)$$
$$\epsilon_\epsilon(y)\epsilon_\beta(x) = \epsilon_\beta(x)\epsilon_\epsilon(y)\epsilon_\zeta(ixy)$$
$$\epsilon_\delta(y)\epsilon_\gamma(x) = \epsilon_\gamma(x)\epsilon_\delta(y)\epsilon_\zeta(jxy)$$

(It is shown in [10] that $a = b = c = d = 1, e = 2, f = g = h = j = 3, i = -1$.) The following lemma, which summarizes the products in G_2 that will be needed, follows directly from the definitions just given.

LEMMA 7.1. $\epsilon_\alpha(x_\alpha)\epsilon_\beta(x_\beta)\epsilon_\gamma(x_\gamma)\epsilon_\delta(x_\delta)\epsilon_\epsilon(x_\epsilon)\epsilon_\zeta(x_\zeta)\epsilon_\alpha(y_\alpha) =$

$$\epsilon_\alpha(x_\alpha+y_\alpha)\epsilon_\beta(x_\beta)\epsilon_\gamma(x_\gamma+ay_\alpha x_\beta)\epsilon_\delta(x_\delta+by_\alpha^2x_\beta+ey_\alpha x_\gamma)\epsilon_\epsilon(z_\epsilon)\epsilon_\zeta(z_\zeta)$$

$$(z_\epsilon = x_\epsilon + cy_\alpha^3x_\beta + fy_\alpha^2x_\gamma + hy_\alpha x_\delta)$$
$$(z_\zeta = x_\zeta + dy_\alpha^3x_\beta^2 + gy_\alpha x_\gamma^2 + bjy_\alpha^2x_\beta x_\gamma)$$
$$\epsilon_\alpha(x_\alpha)\epsilon_\beta(x_\beta)\epsilon_\gamma(x_\gamma)\epsilon_\delta(x_\delta)\epsilon_\epsilon(x_\epsilon)\epsilon_\zeta(x_\zeta)\epsilon_\beta(y_\beta) =$$
$$\epsilon_\alpha(x_\alpha)\epsilon_\beta(x_\beta+y_\beta)\epsilon_\gamma(x_\gamma)\epsilon_\delta(x_\delta)\epsilon_\epsilon(x_\epsilon)\epsilon_\zeta(x_\zeta+ix_\epsilon y_\beta).$$

$$\epsilon_\alpha(x_\alpha)\epsilon_\beta(x_\beta)\epsilon_\gamma(x_\gamma)\epsilon_\delta(x_\delta)\epsilon_\epsilon(x_\epsilon)\epsilon_\zeta(x_\zeta)\epsilon_\gamma(y_\gamma) =$$
$$\epsilon_\alpha(x_\alpha)\epsilon_\beta(x_\beta)\epsilon_\gamma(x_\gamma + y_\gamma)\epsilon_\delta(x_\delta)\epsilon_\epsilon(x_\epsilon)\epsilon_\zeta(x_\zeta + jy_\gamma x_\delta).$$

PROOF. A calculation.

Now define m_1, m_2, m_3 and $\omega_1, \omega_2, \omega_3$ and m_δ, m_ϵ, m_α by the conditions $B\omega_i n = B\omega_i m_i$, $m_3 = \epsilon_\delta(m_\delta)m_2$, $m_2 = \epsilon_\epsilon(m_\epsilon)m_1$, $m_1 = \epsilon_\alpha(m_\alpha)$, $\omega_3 = \sigma_\alpha\omega_2$, $\omega_2 = \sigma_\beta\omega_1$, $\omega_1 = \sigma_\alpha$. Define m_1', m_2', m_3' and ω_1', ω_2', ω_3' and m_β, m_γ, m_ζ by the conditions $B\omega_i' n = B\omega_i' m_i'$, $m_3' = \epsilon_\zeta(m_\zeta)m_2'$, $m_2' = \epsilon_\gamma(m_\gamma)m_1'$, $m_1' = \epsilon_\beta(m_\beta)$, $\omega_3' = \sigma_\beta\omega_2'$, $\omega_2' = \sigma_\alpha\omega_1'$, $\omega_1' = \sigma_\beta$. Also let

$$n = \epsilon_\alpha(n_\alpha)\epsilon_\beta(n_\beta)\epsilon_\gamma(n_\gamma)\epsilon_\delta(n_\delta)\epsilon_\epsilon(n_\epsilon)\epsilon_\zeta(n_\zeta).$$

Lemma 6.1 will be applied six times in what follows – once for each positive root. The group element m of the lemma will take the values m_1, m_2, m_3, m_1', m_2', m_3'. The variable m_γ of the lemma will take on the values $m_\delta, m_\epsilon, m_\alpha, m_\zeta, m_\gamma$, and m_β defined here in successive applications of the lemma. The notation is potentially confusing. Note that the root γ in the lemma need not be the root $\gamma = \alpha + \beta$ of the group G_2, and the simple root α of lemma 6.1 need not be the short simple root of G_2.

This lemma shows how to convert from proposition 6.1 to the given ordering on the roots.

LEMMA 7.2. *With these definitions,*

$$\begin{aligned}
m_\alpha &= n_\alpha \\
m_\beta &= n_\beta \\
m_\gamma &= n_\gamma \\
m_\delta &= n_\delta + bn_\alpha^2 n_\beta - en_\alpha n_\gamma \\
m_\epsilon &= n_\epsilon - cn_\alpha^3 n_\beta + fn_\alpha^2 n_\gamma - hn_\alpha n_\delta \\
m_\zeta &= n_\zeta - in_\beta n_\epsilon - jn_\gamma n_\delta
\end{aligned}$$

(or equivalently)

$$\begin{aligned}
n_\delta &= m_\delta - bm_\alpha^2 m_\beta + em_\alpha m_\gamma \\
n_\epsilon &= m_\epsilon + (c - hb)m_\alpha^3 m_\beta + (he - f)m_\alpha^2 m_\gamma + hm_\alpha m_\delta \\
n_\zeta &= m_\zeta + i(c - hb)m_\alpha^3 m_\beta^2 + (ihe - if - bj)m_\alpha^2 m_\beta m_\gamma + jem_\alpha m_\gamma^2 \\
&\quad + him_\alpha m_\beta m_\delta + jm_\gamma m_\delta + im_\beta m_\epsilon.
\end{aligned}$$

PROOF. The expressions given for the n_-'s in terms of the m_-'s are consequences of the expressions for the m_-'s in terms of the n_-'s. So it is enough to establish the expressions for the m_-'s. By (7.1),

$$nm_1^{-1} = \epsilon_\alpha(n_\alpha - m_\alpha)\epsilon_\beta(n_\beta)\epsilon_\gamma(n_\gamma - am_\alpha n_\beta)\epsilon_\delta(n_\delta + bm_\alpha^2 n_\beta - em_\alpha n_\gamma).$$

$$\text{times} \quad \epsilon_\epsilon(n_\epsilon - cm_\alpha^3 n_\beta + fm_\alpha^2 n_\gamma - hm_\alpha n_\delta)\epsilon_\zeta(*)$$

So $\sigma_\alpha n m_1^{-1} \sigma_\alpha^{-1} \in B$ implies $m_\alpha = n_\alpha$.

$$\begin{aligned} nm_2^{-1} =& nm_1^{-1}\epsilon_\epsilon(-m_\epsilon) \\ =& \epsilon_\beta(n_\beta)\epsilon_\gamma(n_\gamma - am_\alpha n_\beta)\epsilon_\delta(n_\delta + bm_\alpha^2 n_\beta - em_\alpha n_\gamma) \text{ times} \\ & \epsilon_\epsilon(n_\epsilon - m_\epsilon - cm_\alpha^3 n_\beta + fm_\alpha^2 n_\gamma - hm_\alpha n_\delta)\epsilon_\zeta(*). \end{aligned}$$

So $\omega_2 n m_2^{-1} \omega_2^{-1} \in B$ implies
$m_\epsilon = n_\epsilon - cm_\alpha^3 n_\beta + fm_\alpha^2 n_\gamma - hm_\alpha n_\delta$.

$$\begin{aligned} nm_3^{-1} =& nm_2^{-1}\epsilon_\delta(-m_\delta) \\ =& \epsilon_\beta(n_\beta)\epsilon_\gamma(n_\gamma - am_\alpha n_\beta)\epsilon_\delta(n_\delta - m_\delta + bm_\alpha^2 n_\beta - em_\alpha n_\gamma)\epsilon_\zeta(*). \end{aligned}$$

So $\omega_3 n m_3^{-1} \omega_3^{-1} \in B$ implies $m_\delta = n_\delta + bm_\alpha^2 n_\beta - em_\alpha n_\gamma$.

$$nm_1'^{-1} = \epsilon_\alpha(n_\alpha)\epsilon_\beta(n_\beta - m_\beta)\epsilon_\gamma(n_\gamma)\epsilon_\delta(n_\delta)\epsilon_\epsilon(n_\epsilon)\epsilon_\zeta(n_\zeta - im_\beta n_\epsilon).$$

So $\sigma_\beta n m_1^{-1} \sigma_\beta^{-1} \in B$ implies $m_\beta = n_\beta$.

$$\begin{aligned} nm_2'^{-1} =& nm_1'^{-1}\epsilon_\gamma(-m_\gamma) \\ =& \epsilon_\alpha(n_\alpha)\epsilon_\gamma(n_\gamma - m_\gamma)\epsilon_\delta(n_\delta)\epsilon_\epsilon(n_\epsilon)\epsilon_\zeta(n_\zeta - in_\beta n_\epsilon - jm_\gamma n_\delta). \end{aligned}$$

So $\omega_2' n m_2'^{-1} \omega_2'^{-1} \in B$ implies $m_\gamma = n_\gamma$.

$$nm_3'^{-1} = nm_2'^{-1}\epsilon_\zeta(-m_\zeta) = \epsilon_\alpha(n_\alpha)\epsilon_\delta(n_\delta)\epsilon_\epsilon(n_\epsilon)\epsilon_\zeta(n_\zeta - in_\beta n_\epsilon - jm_\gamma n_\delta - m_\zeta).$$

So $\omega_3' n m_3'^{-1} \omega_3'^{-1} \in B$ implies $m_\zeta = n_\zeta - in_\beta n_\epsilon - jm_\gamma n_\delta$.

The main result of this section is the following proposition.

PROPOSITION 7.3. *The coefficients m_γ of G_2 are given as follows.*

$$\begin{aligned} m_\alpha &= 1/z(W_+, \alpha), \\ m_\beta &= 1/z(W_+, \beta), \\ m_{\alpha+\beta} &= 1/z(W_+, \beta)z(W(\sigma_\beta), \alpha), \\ m_{2\alpha+\beta} &= (1/z(W_+, \alpha)z(W(\sigma_\beta\sigma_\alpha), \alpha)z(W(\sigma_\alpha), \beta)) \\ &\quad + (1/z(W_+, \alpha)^2 z(W(\sigma_\alpha), \beta)) \\ m_{3\alpha+\beta} &= -1/z(W_+, \alpha)^3 z(W(\sigma_\alpha), \beta) \\ m_{3\alpha+2\beta} &= -(1/z(W_+, \beta)z(W(\sigma_\alpha\sigma_\beta), \beta)z(W(\sigma_\beta), \alpha)^3) \\ &\quad - (1/z(W_+, \beta)^2 z(W(\sigma_\beta), \alpha)^3). \end{aligned}$$

PROOF. Write $t_3v_3\omega_3m_3 = n_{w_3}; t_2v_2\omega_2m_2 = n_{w_2}; t_1v_1\omega_1m_1 = n_{w_1}$. Let the variables $z_1, z_2, z_3, z_1', z_2', z_3'$ be given by the following diagram.

$$z_1 = z(W_+, \alpha) \qquad z_2 = z(W(\sigma_\alpha), \beta) \qquad z_3 = z(W(\sigma_\beta\sigma_\alpha), \alpha)$$
$$z'_1 = z(W_+, \beta) \qquad z'_2 = z(W(\sigma_\beta), \alpha) \qquad z'_3 = z(W(\sigma_\alpha\sigma_\beta), \beta)$$

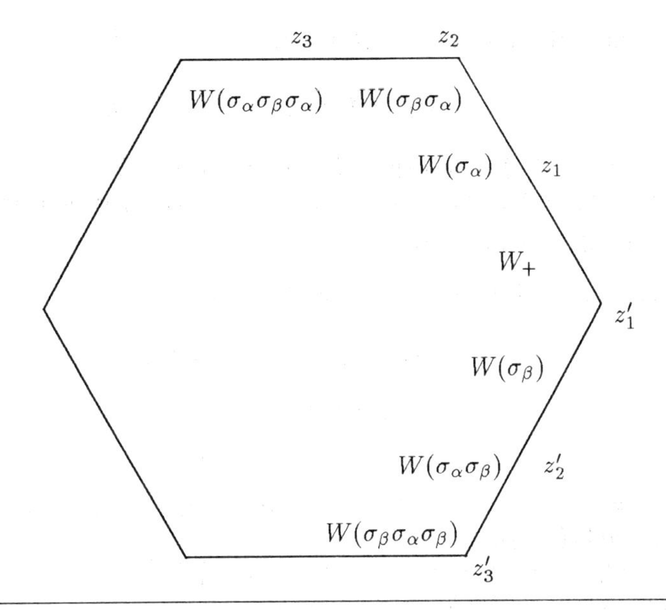

Write

$$n_{w_1} = \exp(z_1 X_{-\alpha}),$$
$$n_{w_2} = \exp(z_2 X_{-\beta}) \exp(z_1 X_{-\alpha}),$$
$$n_{w_3} = \exp(z_3 X_{-\alpha}) \exp(z_2 X_{-\beta}) \exp(z_1 X_{-\alpha}).$$

We now apply lemma 6.1 three times. The *first* application gives

$$\alpha(t_2) z_3 (y_3 - u_3 m_\delta) + 1 = 0$$

where

$$v_2 = \exp(y_3 X_\alpha)^\alpha v_2 \text{ and } \omega_2 X_\delta \omega_2^{-1} = u_3 X_\alpha.$$

The *second* application gives

$$\beta(t_1) z_2 (y_2 - u_2 m_\epsilon) + 1 = 0$$

where

$$v_1 = \exp(y_2 X_\beta)^\beta v_1 \text{ and } \omega_1 X_\epsilon \omega_1^{-1} = u_2 X_\beta;$$
$$t_2 = t_1 (-\beta(t_1) z_2)^{-\beta^v},$$
$$v_2 = \exp(\beta(t_1) z_2 X_\beta) \sigma_\beta \exp(u_2 m_\epsilon X_\beta)^\beta v_1 \exp(-u_2 m_\epsilon X_\beta) \sigma_\beta^{-1}.$$

The *third* application gives

$$z_1 (y_1 - u_1 m_\alpha) + 1 = 0$$

where

$$y_1 = 0 \text{ and } u_1 = 1$$
$$t_1 = (-z_1)^{-\alpha^v},$$
$$v_1 = \exp(z_1 X_\alpha).$$

Straightforward calculations give $\beta(t_1) = \beta(-z_1^{-\alpha^v}) = -z_1^3$,

$$\alpha(t_2) = \alpha(t_1)(-\beta(t_1)z_2)^{-<\alpha,\beta^v>} = z_1 z_2,$$

$$y_1 = 0,\ v_1 = \exp(y_2 X_\beta)^\beta v_1 = \exp(z_1 X_\alpha),\ y_2 = 0,\ \exp(z_1 X_\alpha) = {}^\beta v_1.$$
$$v_2 = \exp(y_3 X_\alpha)^\alpha v_2 = \epsilon_\beta(*)\sigma_\beta \epsilon_\alpha(z_1)\epsilon_\gamma(a z_1 u_2 m_\epsilon)\epsilon_\delta(*)\epsilon_\zeta(*)\sigma_\beta^{-1}$$
$$= \epsilon_\alpha(a z_1 u_2 m_\epsilon u_2')^\alpha v_2.$$

So $y_3 = a z_1 u_2 u_2' m_\epsilon$ where Ad $\sigma_\beta(X_\gamma) = u_2' X_\alpha$. So

$$z_1 z_2 z_3 (a u_2 u_2' m_\epsilon z_1 - u_3 m_\delta) + 1 = 0$$
$$-z_1^3 z_2(-u_2 m_\epsilon) + 1 = 0$$
$$-z_1(-u_1 m_\alpha) + 1 = 0$$

Thus $u_1 m_\alpha = 1/z_1, u_2 m_\epsilon = -1/(z_1^3 z_2), u_3 m_\delta = 1/(z_1 z_2 z_3) + (a u_2 u_2' m_\epsilon z_1) = 1/(z_1 z_2 z_3) - a u_2'/(z_1^2 z_2)$.

Now write $t_3' v_3' \omega_3' m_3' = n_{w_3'}$; $t_2' v_2' \omega_2' m_2' = n_{w_2'}; t_1' v_1' \omega_1' m_1' = n_{w_1'}$. Write $n_{w_1'} = \exp(z_1' X_{-\beta})$, $n_{w_2'} = \exp(z_2' X_{-\alpha}) \exp(z_1' X_{-\beta})$,

$$n_{w_3'} = \exp(z_3' X_{-\beta}) \exp(z_2' X_{-\alpha}) \exp(z_1' X_{-\beta}).$$

We now apply lemma 6.1 three more times. The *first* application gives

$$\beta(t_2') z_3' (y_3' - u_3' m_\zeta) + 1 = 0$$

where $v_2' = \exp(y_3' X_\beta)^\beta v_2'$ and $\omega_2' X_\zeta \omega_2'^{-1} = u_3' X_\beta$. The *second* application gives

$$\alpha(t_1') z_2' (y_2' - u_2' m_\gamma) + 1 = 0$$

where

$$v_1' = \exp(y_2' X_\alpha)^\alpha v_1' \text{ and } \omega_1' X_\gamma \omega_1'^{-1} = u_2' X_\alpha;$$
$$t_2' = t_1'(-\alpha(t_1') z_2')^{-\alpha^v},$$
$$v_2' = \exp(\alpha(t_1') z_2' X_\alpha) \sigma_\alpha \exp(u_2' m_\gamma X_\alpha)^\alpha v_1' \exp(-u_2' m_\gamma X_\alpha) \sigma_\alpha^{-1}.$$

The *third* application gives

$$z_1'(y_1' - u_1' m_\beta) + 1 = 0 \text{ where } y_1' = 0 \text{ and } u_1' = 1,$$

$$t_1' = (-z_1')^{-\beta^v},$$
$$v_1' = \exp(z_1' X_\beta).$$

Now we have as a consequence $\alpha(t_1') = \alpha(-z_1'^{-\beta^v}) = -z_1'$,

$$\beta(t_2') = \beta(t_1')(-\alpha(t_1')z_2')^{-<\beta,\alpha^v>} = z_1' z_2'^3,$$

$$y_1' = 0,$$

$$v_1' = \exp(y_2' X_\alpha)^\alpha v_1' = \exp(z_1' X_\beta),$$

$$y_2' = 0, {}^\alpha v_1' = \exp(z_1' X_\beta).$$

$$v_2' = \exp(y_3' X_\beta)^\beta v_2' = \epsilon_\alpha(*)\sigma_\alpha \epsilon_\alpha(u_2' m_\gamma)\epsilon_\beta(z_1')\epsilon_\alpha(-u_2' m_\gamma)\sigma_\alpha^{-1} =$$

$$\epsilon_\alpha(*)\sigma_\alpha \epsilon_\beta(z_1')\epsilon_\gamma(*)\epsilon_\delta(*)\epsilon_\epsilon(c(-u_2' m_\gamma)^3 z_1')\epsilon_\zeta(*)\sigma_{\alpha^{-1}} =$$
$$\epsilon_\beta(c(-u_2' m_\gamma)^3 z_1' u_2)^\beta v_2'.$$

So

$$y_3' = -c(u_2' m_\gamma)^3 z_1' u_2$$

where

$$\mathrm{Ad}\ \sigma_\alpha(X_\epsilon) = u_2 X_\beta.$$

This gives the equations

$$-z_1' u_1' m_\beta + 1 = 0 \text{ or } u_1' m_\beta = 1/z_1'$$

$$z_1' z_2' u_2' m_\gamma + 1 = 0 \text{ or } u_2' m_\gamma = -1/(z_1' z_2')$$

$$z_1' z_2'^3 z_3'(-c u_2'^3 m_\gamma^3 z_1' u_2 - u_3' m_\zeta) + 1 = 0$$

$$\text{or} \quad u_3' m_\zeta = 1/(z_1' z_2'^3 z_3') - c(u_2' m_\gamma)^3 z_1' u_2 =$$
$$= 1/(z_1' z_2'^3 z_3') + c u_2/(z_1'^2 z_2'^3).$$

Humphreys [10] shows that Ad $\sigma_\alpha(X_\delta) = -X_\gamma$, Ad $\sigma_\alpha(X_\epsilon) = X_\beta$, Ad $\sigma_\alpha(X_\zeta) = X_\zeta$, Ad $\sigma_\beta(X_\gamma) = -X_\alpha$, Ad $\sigma_\beta(X_\delta) = X_\delta$, Ad $\sigma_\beta(X_\zeta) = -X_\epsilon$. From this it follows that $u_1 = 1, u_2 = 1, u_3 = 1, u_1' = 1, u_2' = -1, u_3' = -1$. Substituting these values into the expressions for m_-, we obtain the result.

8. The Functions n_γ.

The functions n_γ are much easier to compute for the classical groups. In fact lemma 6.1 is not needed. To illustrate we calculate them for $SL(n)$. (The same method gives all classical groups).

LEMMA 8.1.

$$n_{\alpha_r + \dots + \alpha_{r+s}} = 1/(z(W_0, \alpha_r) z(W_1, \alpha_{r+1}) \dots z(W_s, \alpha_{r+s}))$$

where $W_0 = W_+$, and W_{t+1} is adjacent to W_t through a wall of type $\alpha_{r+t}(t = 0, \dots, s-1)$.

PROOF. Represent $SL(n)$ in the standard way with $\mathbb{B} = B_0$ upper triangular and B_∞ lower triangular. The relation $\mathbb{B}^{wn} = \mathbb{B}^{n_w}$ is equivalent to $wnn_w^{-1} \in \mathbb{B}$.

Let $\sigma_i = \sigma_{\alpha_{r+i}}, X_i = X_{-\alpha_{r+i}}, z_i = z(W_i, \alpha_{r+i}), \epsilon_i(x) = \exp(xX_i)$, $i = 0, \ldots, s$. Then $k_t =^{(def)} \omega n n_w^{-1} =$

$$\sigma_t \ldots \sigma_0 n \epsilon_0(-z_0) \ldots \epsilon_t(-z_t) \in B.$$

Also

$$\begin{aligned} k_{t+1} &= \sigma_{t+1}\sigma_t \ldots \sigma_0 n \epsilon_0(-z_0) \ldots \epsilon_t(-z_t)\epsilon_{t+1}(-z_{t+1}) \\ &= \sigma_{t+1} k_t \epsilon_{t+1}(-z_{t+1}) \in B. \end{aligned}$$

Since σ_{t+1} and $\epsilon_{t+1}(-z_{t+1})$ belong to the rank one subgroup with Lie algebra $X_\alpha, X_{-\alpha}, H_\alpha$ where $\alpha = \alpha_{r+t+1}$ we can compute this last product inside $SL(2)$ provided that we can determine the coefficients of k_t in the $c_{i,i}, c_{i,i+1}, c_{i+1,i}$ and $c_{i+1,i+1}$ positions where $i = r + t + 1$. Since $k_t \in B$, $c_{i+1,i} = 0$. σ_i acts as the permutation $(r + i, r + i + 1)$ on the rows of nn_w^{-1}. So $\sigma_t \ldots \sigma_0$ acts as the permutation $(r + t, r + t + 1) \ldots (r, r + 1) = (r + t + 1, r + t, \ldots, r)$ on the rows of nn_w^{-1}. Thus the $r + t + 1st$ row of $\omega n n_w^{-1}$ equals the rth row of nn_w^{-1} and the $r + t + 2nd$ row of $\omega n n_w^{-1}$ equals the $r + t + 2nd$ row of nn_w^{-1}. Since $n_w^{-1} = \epsilon_0(-z_0) \ldots \epsilon_t(-z_t)$ the ith and $i+1$st columns of nn_w^{-1} are the same as the respective columns of n. Thus $c_{i+1,i+1} = 1$, $c_{i,i+1} = n_{r,r+t+2} = n_{\alpha_r + \ldots + \alpha_{r+t+1}}$, $c_{i,i} = n_{r,r+t+1} = n_{\alpha_r + \ldots + \alpha_{r+t}}$.

We have the matrix product in SL(2)

$$\begin{pmatrix} 0 & 1 \\ -1 & 0 \end{pmatrix} \begin{pmatrix} x & y \\ 0 & 1 \end{pmatrix} \begin{pmatrix} 1 & 0 \\ -z & 1 \end{pmatrix} = \begin{pmatrix} -z & 1 \\ yz - x & -y \end{pmatrix} \in B.$$

The equality $x = yz$ becomes $c_{i,i} = z_{t+1} c_{i,i+1}$.

$$n_{\alpha_r + \ldots + \alpha_{r+t}} / z(W_{t+1}, \alpha_{r+t+1}) = n_{\alpha_r + \ldots + \alpha_{r+t+1}},$$

so by induction

$$n_{\alpha_r + \ldots + \alpha_{r+s}} = 1/(z(W_0, \alpha_r) z(W_1, \alpha_{r+1}) \ldots z(W_s, \alpha_{r+s})).$$

Also by a similar argument the following lemma is proved.

LEMMA 8.2. *Let m_β be the βth coefficient of the inverse of n, let $W_0 = W_+$ and let W_{j+1} be the Weyl chamber adjacent to W_j through the wall (W_j, α_{r-j}) then*

$$m_{\alpha_{r-s} + \ldots + \alpha_r} = (-1)^{s+1} / z(W_0, \alpha_r) z(W_1, \alpha_{r-1}) \ldots z(W_s, \alpha_{r-s}).$$

PROOF. Again represent B_0 in $SL(n)$ as the upper triangular matrices and B_∞ as the lower triangular matrices. The calculation takes place entirely inside a Levi subgroup of P_Σ where P_Σ is the parabolic subgroup given by the simple roots $\Sigma = \{\alpha_{r-s}, \ldots, \alpha_r\}$. Thus we reduce immediately to the case where $r - s = 1$ and $r = n - 1$. Let $\sigma_i = \sigma_{\alpha_i}, X_i = X_{-\alpha_i}$, and let $z_i = z(W_{i-1}, \alpha_{n-i})$. As in the previous lemma we have $\omega n n_w^{-1} \in B_0$.

Let

$$n_w = \exp(z_{n-1}X_1)\dots\exp(z_1 X_{n-1}).$$

Then

$$\sigma_1\dots\sigma_{n-1} n \exp(-z_1 X_{n-1})\dots\exp(-z_{n-1}X_1) \in B.$$

To make it possible to use the previous lemma we apply the involution $g \to J^t g^{-1} J$ to the group $SL(n)$ where

$$J = \begin{pmatrix} 0 & 0 & 0 & 0 & 1 \\ 0 & 0 & 0 & 1 & 0 \\ 0 & 0 & \dots & 0 & 0 \\ 0 & 1 & 0 & 0 & 0 \\ 1 & 0 & 0 & 0 & 0 \end{pmatrix}.$$

Under this involution $X_i \to -X_{n-i}, \sigma_i \to \sigma_{n-i}^{-1}, n \to J^t n^{-1} J$. Thus $\omega n n_w^{-1}$ is sent to

$$\sigma_{n-1}^{-1}\dots\sigma_1^{-1} J^t n^{-1} J \exp(z_1 X_1)\dots\exp(z_{n-1}X \in B.$$

Now the coefficient in the $(1,n)$th position of $J^t n^{-1} J$ equals $m_{\alpha_1+\dots+\alpha_{n-1}}$. We now apply the previous lemma bearing in mind that the signs of the z_i have been reversed. This gives the result.

9. The Fundamental Divisors on Y_Γ.

There is a morphism from each divisor E in Y_Γ to the variety $\xi^{-1}(1)$ in Y_1. An irreducible divisor E in Y_Γ whose image in Y_1 lies in the complement of Y'' is called a *spurious divisor*. If there are walls (W,α) and (W',α) such that $z(W,\alpha)/z(W',\alpha)(= z_1(W,\alpha)/z_1(W',\alpha)) = 0$ on E, then E is a spurious divisor. By [15,3.8] the condition that $z(W,\alpha)/z(W',\alpha) = 0$ is independent of the coordinate patch. All other irreducible divisors on Y_Γ are called *fundamental divisors*. An open dense set of a fundamental divisor lies over Y''.

Every fundamental divisor in Y_Γ maps to (but not necessarily onto) an irreducible component of $\xi^{-1}(1)$ in Y_1. Much can be learned about the fundamental divisors by studying their image in Y_1. In fact nearly all calculations with fundamental divisors can be reduced to calculations in the singular variety Y_1. We use the coordinates developed in this chapter to study the fundamental divisors.

Weil divisors are defined on any variety which is regular in codimension one [7]. I have not proved in general that Y_1'' is regular in codimension one, although it seems likely that it is. Thus whenever I speak of a divisor E on (a subvariety of) Y'', I must first produce an appropriate subvariety of Y'' which is regular in codimension one. Assuming that this can be done, we also call the divisors of $\lambda = 0$ in Y_1'' fundamental divisors. Provided appropriate regularity conditions hold, they are given by the irreducible components of the set of the following set of equations with λ set equal to 0:

$$w(\gamma)\lambda = x(\gamma)\prod z(\alpha)^{m(\alpha)} : \gamma = \sum m(\alpha)\alpha$$

$$w(\gamma)x(\beta) = w(\beta)x(\gamma)\prod z(\alpha)^{m(\alpha)} : \gamma - \beta = \sum m(\alpha)\alpha$$

(The full list of equations here includes all equations holding on the Zariski closure of the variety defined for $\lambda \neq 0$ by $\lambda w(\gamma) = x(\gamma) \prod z(\alpha)^{m(\alpha)} : \gamma = \sum m(\alpha)\alpha$.)

Recall that $w(\alpha) = 1$ for α simple. There are $2N + 1$ variables in the set

$$\{\lambda, w(\gamma), x(\gamma), z(\alpha)\} \backslash \{w(\alpha) : \alpha \text{ simple}\}$$

where $N = (dim(G) - rank(G))/2 = dim(S''(B_\infty, B_0))$. For any positive root γ, set

$$\gamma = \sum m_\gamma(\alpha)\alpha, \quad m_\gamma = \sum m_\gamma(\alpha) \quad \text{and} \quad z(\gamma) = \prod z(\alpha)^{m_\gamma(\alpha)}.$$

For every set Σ of simple roots, we describe a divisor E_Σ (again assuming appropriate regularity). Let Σ' be the positive roots spanned by Σ; and let Σ^c be the other positive roots. E_Σ is the closure in Y'' of the subvariety defined by

i) $\lambda = 0$
ii) $x(\alpha) \neq 0$ for $\alpha \notin \Sigma : \alpha$ simple
iii) $z(\alpha) \neq 0$ for $\alpha \in \Sigma : \alpha$ simple.

PROPOSITION 9.1. *Let Y''_Σ be the open subvariety of Y'' consisting of points on Y'' which satisfy conditions (ii) and (iii). The open subvariety of Y'' consisting of the union of Y''_Σ $\forall\, \Sigma$ is regular in codimension one. E_Σ is an O_Σ-divisor where O_Σ is the Richardson class of the parabolic subgroup P_Σ. The Igusa constant $a(E_\Sigma)$ is one. The Igusa constant $\beta(E_\Sigma) - 1$ equals $(dim(C_G(u) - rank(G))/2$.*

PROOF. The local ring near a generic point of E_Σ is regular because it is generated by $\{\lambda, x(\gamma) : \gamma \in \Sigma^c, w(\gamma) : \gamma \in \Sigma', z(\alpha) : \alpha \in \Sigma\}$ for :

$$z(\alpha) = \lambda / x(\alpha) : \alpha \notin \Sigma;$$

$$x(\gamma) = \lambda w(\gamma)/z(\gamma) : \gamma \in \Sigma';$$

$$w(\gamma) = x(\gamma) \prod z(\alpha)^{m(\alpha)}/x(\alpha_0) : \gamma \in \Sigma^c, \alpha_0 \notin \Sigma, \gamma - \alpha_0 = \sum m(\alpha)\alpha$$

(γ not simple). We see that $x(\gamma) = 0$ on E_Σ if and only if $\gamma \in \Sigma'$. This proves that the subvariety of Y'' of the lemma is regular in codimension one. This also shows that that E_Σ is an O_Σ-divisor where O_Σ is the Richardson class of the parabolic P_Σ. We also see that the Igusa constant $a(E_\Sigma)$, which is defined to be the multiplicity with which λ vanishes along E, equals one.

Spaltenstein [22,3.2] tells us that $\dim C_G(v) = \dim(M_\Sigma)$ where M_Σ is a Levi component associated to the parabolic P_Σ and $v \in O_\Sigma$. The form on X^0 is given by

$$\omega_X = d\lambda \wedge dx(\alpha) \wedge dx(\beta) \wedge \ldots \wedge d\nu_\alpha \wedge d\nu_\beta \wedge \ldots$$

To pass to the form in a neighborhood of E_Σ we make the substitution $x(\gamma) = \lambda w(\gamma)/z(\gamma)$ for $\gamma \in \Sigma'$ where $z(\gamma)$ is defined to be the product $\prod z(\alpha)^{m(\alpha)}$. We can pull the factors λ out of the form to obtain

$$\omega_X = \lambda^{|\Sigma'|} d\lambda \wedge \prod d(w(\gamma)/z(\gamma)) \wedge \prod dx(\beta) \ldots = \lambda^{\beta(E)-1} d\lambda \wedge \ldots.$$

The constant $b(E)-1$ is defined to be the multiplicity of the zero of ω_X along E which in this case is $|\Sigma'|$. When $a(E)=1, \beta(E)=b(E)$. But then $2(\beta(E)-1)=2|\Sigma'|=dim(M_\Sigma)-rank(G)$. This completes the proof.

The result that $\beta(E)-1=(dim(C_G(u))-rank(G))/2$ is to be expected. Harish-Chandra has proved the following result in characteristic zero.

PROPOSITION 9.2. *Let O be an F-unipotent conjugacy class. Let $\Gamma(\gamma)$ be the O-germ of an orbital integral relative to the measure*

$$\prod(1-\alpha^{-1}(\gamma))\omega_{T\backslash G}.$$

Let $r=dim(C_G(u))-rank(G)$. Let X be a vector in the Lie algebra of T which does not lie in any singular hyperplane. Then $\Gamma(\exp(\lambda^2X))=|\lambda|^r\Gamma(\exp(X))$.

PROOF. [5]

Let E_1 denote a globally defined divisor which is equal to E_Σ on the given coordinate patch $Y''(B_\infty,B_0)$. The following lemma insures that Σ is independent of the coordinate patch and that divisors which are distinct on one coordinate patch are distinct on every coordinate patch.

LEMMA 9.3. *Let $(u,B(W))$ be a generic point of E_1. Let $P_{\Sigma'}$ be a parabolic subgroup minimal among those parabolic subgroups containing $B(W)$ for all W. Then $\Sigma=\Sigma'$.*

PROOF. On a coordinate patch $Y''(B_\infty,B_0)$, we have seen that u is a Richardson class of the parabolic subgroup P_Σ. Also $z(\alpha)=0,\alpha\notin\Sigma$. This implies that $B(W)\in P_\Sigma$ for all W. E_1 is the closure of E_Σ, so that $B(W)\in P_\Sigma$ for all W and all points in E_1. Since $z(W_+,\alpha)\neq 0$ for $\alpha\in\Sigma$ at the generic point, we have $B(W_+)\neq B(W(\sigma_\alpha))$, so that $\alpha\in\Sigma'$ for any $P_{\Sigma'}$ containing $B(W)$ for all W. Thus $\Sigma\subseteq\Sigma'$, and P_Σ is minimal.

The following simple fact will be needed in the proof of proposition 9.5. We continue to work over the algebraic closure $\bar{F}$ of F.

LEMMA 9.4. *Every irreducible component of $\lambda=0$ has codimension one in Y''.*

PROOF. This follows directly from [18,p.65]

The following proposition will not be needed in what follows.

PROPOSITION 9.5. *When $G=A_n$, the variety Y'' is regular in codimension one. The only divisors on Y'' for a group of type A_n are E_Σ $\forall$ Σ.*

PROOF. If we show that every point of $\lambda=0$ lies inside one of the divisors E_Σ, then the result follows from proposition 9.1. Let E be a component of $\lambda=0$. We will use the following observation repeatedly. If E has dimension x and the coordinate ring of E is generated by $x+r$ functions then at most r of those functions equal zero.

Let R be the set of roots γ such that $x(\gamma) \neq 0$ on E. Let $R_{\min}$ be the subset of R such that if $\gamma \in R_{\min}$ and $\gamma - \beta = \sum m(\alpha)\alpha$ with $m(\alpha) \geq 0$ then $x(\beta) = 0$. Since $w(\alpha) = 1$ for α simple, the functions $\{w(\gamma)\}$ are indexed by $\gamma \in \Phi^+\backslash\Delta$ where Φ^+ are the positive roots and Δ are the simple roots. Then there are $|\{\lambda\}| + |\{w(\gamma)\}| + |\{z(\alpha)\}| + |\{x(\gamma)\}| + |\{\text{ coefficients of } \nu\}| = 1 + (N - \ell) + \ell + N + N = 3N + 1$ variables where $\ell = rank(G)$ and $2N = (dim(G) - \ell)$. Also by (9.4) the dimension of E equals $2N$. The observation above tells us that the number of variables we eliminate plus the number that are identically zero is at most $N + 1$.

We know by definition of R that $x(\beta) = 0$ on E for $\beta \in \Phi^+\backslash R$. If $\gamma \in R\backslash R_{\min}$ then there exists a positive root β with $x(\beta) \neq 0$ such that $\gamma - \beta = \sum m(\alpha)\alpha$, $m(\alpha) \geq 0$. For such γ the variable $w(\gamma)$ can be eliminated through the equation

$$w(\gamma)x(\beta) = x(\gamma)w(\beta) \prod z(\alpha)^{m(\alpha)}.$$

We know that that $\lambda = 0$ on E. If $\alpha \in R_{\min} \cap \Delta$ then the equation $z(\alpha)x(\alpha) = \lambda$ allows one to eliminate $z(\alpha)$. In summary if $\delta = |R_{\min}\backslash\Delta|$ then we either eliminate or set to zero $(N - |R|) + |R\backslash R_{\min}| + 1 + |R_{\min} \cap \Delta| = N + 1 - \delta$ variables.

Now specialize to the case $G = A_n$. We can write the elements of $R_{\min}$ as $\gamma_1, \ldots, \gamma_p$ where $\gamma_i = \alpha_{r_i} + \cdots + \alpha_{r_i + s_i}$ with $r_1 < \cdots < r_p$ and $r_1 + s_1 < \cdots < r_p + s_p$ and $s_i \geq 0$.

LEMMA 9.6. $R_{\min} \subseteq \Delta$.

PROOF. For each $\gamma \in R_{\min}\backslash\Delta$, write $w(\gamma)x(\beta) = w(\beta)x(\gamma)z(\alpha_r)$ where $\gamma = \alpha_r + \ldots + \alpha_{r+s}$ and $\beta = \gamma - \alpha_r$. By the definition of $R_{\min}$, $x(\beta) = 0$ and $x(\gamma) \neq 0$. So $w(\beta)z(\alpha_r) = 0$. From the the inequalities $r_1 < \cdots < r_p$ it follows that the variables $z(\alpha_r)$ are distinct and do not equal any of the variables $z(\alpha)(\alpha \in R_{\min} \cap \Delta)$ eliminated in the previous paragraph. From the same inequalities it follows that the variables $w(\beta)$ are distinct. Furthermore β cannot lie in $R\backslash R_{\min}$ for this would force γ to lie in $R\backslash R_{\min}$ as well. Thus the equations $w(\gamma - \alpha_r)z(\alpha_r) = 0 \;\forall\; \gamma \in R_{\min}\backslash\Delta$ force $\delta = |R_{\min}\backslash\Delta|$ addition variables to zero. We have now eliminated or set to zero $N+1$ distinct variables. No further variables can be eliminated or set to zero without contradiction.

Set $w_i = w(\alpha_{r_i} + \cdots + \alpha_{r_i + s_i - 1})$ and $z_i = z(\alpha_{r_i + s_i})$ for $i \in \{1, \ldots, p\}$. Let q be the largest element of $\{1, \ldots, p\}$ such that $\gamma_q \in R_{\min}\backslash\Delta$. As above we have $w_i z_i = 0$. Suppose $z_q = 0$, then we must have $z_q = z(\alpha_r)$ for some r. But since $r_{q-1} < r_q < r_q + s_q < r_{q+1} + s_{q+1} = r_{q+1}$, this is impossible. Thus $z_q \neq 0$ and $w_q = 0$. Continuing inductively suppose that $z_{j+1}, \ldots, z_q \neq 0$ and $\gamma_j, \ldots, \gamma_q \in R_{\min}\backslash\Delta$. Then it follows that again $z_j \neq 0$ and $w_j = 0$. If j is now chosen to be the smallest integer such that $\gamma_j, \ldots, \gamma_q \in R_{\min}\backslash\Delta$ then $w_j = 0$. Since $\gamma_{j-1} \in \Delta$, w_j cannot equal any of the variables previously set to zero. This gives a contradiction. Thus $R_{\min} \subseteq \Delta$.

Since no variables equal zero on E other than those already specified, we have on E (i) $\lambda = 0$, (ii) $x(\alpha) \neq 0$ for $\alpha \notin \Delta\backslash R_{\min}$ (that is $\alpha \in R_{\min}$), (iii) $z(\alpha) \neq 0$ for $\alpha \in \Delta\backslash R_{\min}$. These are the conditions holding on an open set of the divisor E_Σ with $\Sigma = \Delta\backslash R_{\min}$.

III. GROUPS OF RANK TWO

1. Zero Patterns.

Consider the variety of stars S_1 and a divisor E. For every coordinate patch $S_1(B_\infty, B_0)$ and simple root α select a chamber W_α such that $z_1(W_\alpha, \alpha) \neq 0$ on E. Certain of the variables $z(W,\alpha)/z(W_\alpha,\alpha)(= z_1(W,\alpha)/z_1(W_\alpha,\alpha))$ will vanish identically on E. This gives a *zero pattern*, i.e., a map θ_1 from the walls to $\{0,1\}$, by $\theta_1(W,\alpha) = 0$ if and only if $z(W,\alpha)/z(W_\alpha,\alpha) = 0$ on E. The zero pattern depends only on E and not on the choices (B_∞, B_0) and W_α. The first part of this chapter studies the zero patterns for groups of rank two.

The rank two zero patterns will then be used to prove a result about the divisors for an arbitrary group. Let θ_1 be a zero pattern. For a fixed point p (which is not necessarily a closed point of the variety), a *special node* is a node such that at least one wall of each type is non-zero but that at least one wall is zero. By the nature of the equations at a node of type $A_1 \times A_1$,

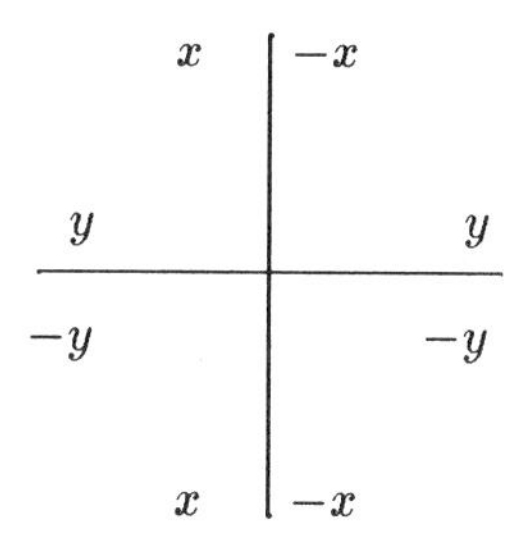

a special node must be of type A_2, B_2, or G_2. The following proposition indicates the importance of special nodes.

PROPOSITION 1.1. *Let G be a group of semi-simple rank 2. Suppose that at a generic point of a divisor E there is a special node. Then E makes no contribution to the subregular germ.*

The proof of this proposition is the subject of the second half of this chapter. Fix a torus T and a Borel B of G and let the positive roots be

$$\alpha, \beta, \gamma = \alpha + \beta \qquad A_2$$
$$\alpha, \beta, \gamma = \alpha + \beta,\ \delta = 2\alpha + \beta \qquad B_2$$
$$\alpha, \beta, \gamma = \alpha + \beta, \delta = 2\alpha + \beta, \epsilon = 3\alpha + \beta, \zeta = 3\alpha + 2\beta \qquad G_2.$$

We will take products according to this order on the roots.

LEMMA 1.2. (G_2)
$\epsilon_\alpha(x_1)\epsilon_\beta(y_1)\epsilon_\alpha(x_2)\epsilon_\beta(y_2)\ldots\epsilon_\alpha(x_n)\epsilon_\beta(y_n) =$
$\epsilon_\alpha(a_n)\epsilon_\beta(b_n)\epsilon_{\alpha+\beta}(c_n)\epsilon_{2\alpha+\beta}(d_n)\epsilon_{3\alpha+\beta}(e_n)\epsilon_{3\alpha+2\beta}(f_n)$
where

$$
\begin{aligned}
a_n &= x_1 + x_2 + \ldots + x_n \\
b_n &= y_1 + y_2 + \ldots + y_n \\
c_n &= (x_2 + \ldots + x_n)y_1 + (x_3 + \ldots + x_n)y_2 + \ldots + x_n y_{n-1} \\
d_n &= (x_2 + \ldots + x_n)^2 y_1 + (x_3 + \ldots + x_n)^2 y_2 + \ldots + x_n^2 y_{n-1} \\
e_n &= (x_2 + \ldots + x_n)^3 y_1 + (x_3 + \ldots + x_n)^3 y_2 + \ldots + x_n^3 y_{n-1}
\end{aligned}
$$

PROOF. For $n = 1$ the statement is obvious. We proceed by induction. By lemma II.7.1,
$\epsilon_\alpha(a_{n-1})\epsilon_\beta(b_{n-1})\epsilon_\gamma(c_{n-1})\epsilon_\delta(d_{n-1})\epsilon_\epsilon(e_{n-1})\epsilon_\varsigma(f_{n-1})\epsilon_\alpha(x_n)\epsilon_\beta(y_n)$
$= \epsilon_\alpha(a_{n-1} + x_n)\epsilon_\beta(b_{n-1})\epsilon_\gamma(b_{n-1}x_n + c_{n-1})\epsilon_\delta(d_{n-1} + 2x_n c_{n-1} + x_n^2 b_{n-1})$
$\epsilon_\epsilon(e_{n-1} + 3x_n d_{n-1} + 3x_n^2 c_{n-1} + x_n^3 b_{n-1})\epsilon_\varsigma(*)\epsilon_\beta(y_n)$
$= \epsilon_\alpha(a_{n-1} + x_n)\epsilon_\beta(b_{n-1} + y_n)\epsilon_\gamma(b_{n-1}x_n + c_{n-1})\epsilon_\delta(d_{n-1} + 2x_n c_{n-1} + x_n^2 b_{n-1})$
$\epsilon_\epsilon(e_{n-1} + 3x_n d_{n-1} + 3x_n^2 c_{n-1} + x_n^3 b_{n-1})\epsilon_\varsigma(*)$.
Define $p_n(i)$ by

$$
p_n(i) = \begin{cases} \sum_{j=2}^n (x_j + \ldots + x_n)^i y_{j-1} & \text{for } i \geq 1 \\ y_1 + \ldots + y_n & \text{for } i = 0. \end{cases}
$$

It follows by expanding $((x_j + \ldots + x_{n-1}) + x_n)^i$ in powers of x_n that $p_n(i) = \sum_{k=0}^i \binom{i}{k} x_n^{i-k} p_{n-1}(k)$. We wish to show that $b_n = p_n(0)$, $c_n = p_n(1)$, $d_n = p_n(2)$, $e_n = p_n(3)$. By induction we have

$$
\begin{aligned}
c_n &= c_{n-1} + x_n b_{n-1} = p_{n-1}(1) + x_n p_{n-1}(0) \\
&= p_n(1) \\
d_n &= d_{n-1} + 2x_n c_{n-1} + x_n^2 b_{n-1} \\
&= p_{n-1}(2) + 2x_n p_{n-1}(1) + x_n^2 p_{n-1}(0) = \\
&= p_n(2) \\
e_n &= e_{n-1} + 3x_n d_{n-1} + 3x_n^2 c_{n-1} + x_n^3 b_{n-1} \\
&= p_{n-1}(3) + 3x_n p_{n-1}(2) + 3x_n^2 p_{n-1}(1) + x_n^3 p_{n-1}(0) \\
&= p_n(3).
\end{aligned}
$$

LEMMA 1.3. *Suppose*
(*) $\epsilon_\alpha(x_1)\epsilon_\beta(y_1)\epsilon_\alpha(x_2)\ldots\epsilon_\alpha(x_n)\epsilon_\beta(y_n) = 1$, *then*
a) *if* $n = 1, x_1 = y_1 = 0$;
b) *if* $n = 2, x_1 = x_2 = 0$ or $y_1 = y_2 = 0$;
c) (B_2, G_2) *if* $n = 3, x_1 = x_2 = x_3 = 0$, *or* $y_1 = y_2 = y_3 = 0$, *or* $x_i = y_{i+1} = 0$ *for some* i, *where the subscripts are read modulo three.*

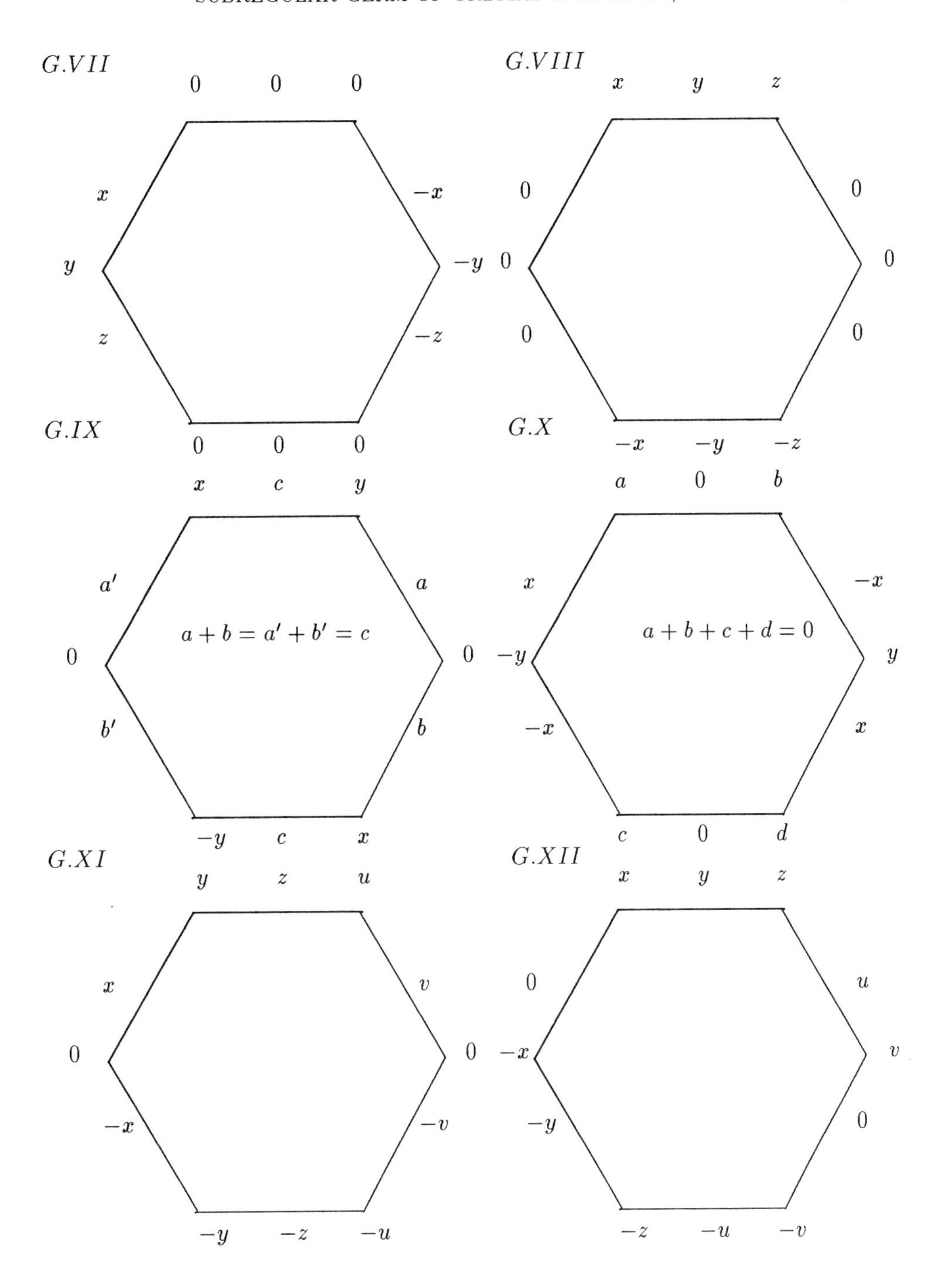

A few remarks on the diagrams are in order. Consider first the case A_2. Suppose $x_2 = 0$ for instance. Then

$$\epsilon_\alpha(x_1)\epsilon_\beta(y_1 + y_2)\epsilon_\alpha(x_3)\epsilon_\beta(y_3) = 0.$$

Lemma 1.3.b together with the fact that at least one variable of each type must be non-zero implies that $y_1 + y_2 = y_3 = 0$.

Consider B_2. Suppose $x_3 = 0$. Then lemma 1.3.c together with the hypothesis that at least one wall of type α is non-zero implies that $y_1 = y_2 + y_3 = y_4 = 0$ (pattern $B.I$), or $x_2 = x_3 = y_4 = 0$ if the x_i of (1.3.c) is adjacent to x_3 (pattern $B.II$), or $x_3 = x_1 = y_2 + y_3 = 0$ if the x_i of the lemma is not adjacent to x_2 (pattern $B.III$). Finally suppose that none of the walls x_i are zero. Then suppose $y_1 = 0$. Again we apply (1.3.c) noting that the first two possibilities do not arise, so that we have $x_1 + x_2 = y_3 = 0$ (pattern $B.IV$).

I will not give a proof that the zero patterns listed above for G_2 are the only possibilities but I will make the results that follow independent of this classification.

2. Coordinate Relations.

As a next step in the proof of proposition 1.1 it is necessary to obtain formulas relating the functions $w(\gamma)$ to the coordinates $z(W, \alpha)$. Lemmas 7.2 and 7.3 relate the coefficients n_β to the coordinates $z(W, \alpha)$. Thus it is enough to calculate the dependence of $w(\gamma)$ on the coefficients n_β. Lemmas 2.1 and 2.2 carry out these calculations.

LEMMA 2.1. *Defining $a, b, c, d, e, f, g, h, i, j$ as in $(II.7)$ we have*

$$\epsilon_\zeta(x_\zeta)\epsilon_\epsilon(x_\epsilon)\epsilon_\delta(x_\delta)\epsilon_\gamma(x_\gamma)\epsilon_\beta(x_\beta)\epsilon_\alpha(x_\alpha)$$

times

$$\epsilon_\alpha(y_\alpha)\epsilon_\beta(y_\beta)\epsilon_\gamma(y_\gamma)\epsilon_\delta(y_\delta)\epsilon_\epsilon(y_\epsilon)\epsilon_\zeta(y_\zeta)$$

$$= \epsilon_\alpha(z_\alpha)\epsilon_\beta(z_\beta)\epsilon_\gamma(z_\gamma)\epsilon_\delta(z_\delta)\epsilon_\epsilon(z_\epsilon)\epsilon_\zeta(z_\zeta)$$

with

$$\begin{aligned}
z_\alpha &= x_\alpha + y_\alpha \\
z_\beta &= x_\beta + y_\beta \\
z_\gamma &= x_\gamma + y_\gamma + a(x_\alpha + y_\alpha)x_\beta \\
z_\delta &= x_\delta + y_\delta + b(x_\alpha + y_\alpha)^2 x_\beta + e(x_\alpha + y_\alpha)x_\gamma \\
z_\epsilon &= x_\epsilon + y_\epsilon + c(x_\alpha + y_\alpha)^3 x_\beta + f(x_\alpha + y_\alpha)^2 x_\gamma + h(x_\alpha + y_\alpha)x_\delta \\
z_\zeta &= x_\zeta + y_\zeta + d(x_\alpha + y_\alpha)^3 x_\beta^2 + g(x_\alpha + y_\alpha)x_\gamma^2 + i(x_\beta + y_\beta)x_\epsilon + j(x_\gamma + y_\gamma)x_\delta \\
&\quad + hi(x_\alpha + y_\alpha)(x_\beta + y_\beta)x_\delta + fi(x_\alpha + y_\alpha)^2(x_\beta + y_\beta)x_\gamma + ci(x_\alpha + y_\alpha)^3 x_\beta y_\beta \\
&\quad + bj(x_\alpha + y_\alpha)^2 x_\beta y_\gamma + ej(x_\alpha + y_\alpha)x_\gamma y_\gamma + aj(x_\alpha + y_\alpha)x_\beta x_\delta + aje(x_\alpha + y_\alpha)^2 x_\beta x_\gamma .
\end{aligned}$$

PROOF. The underlined quantities are moved to the left in each step.

$\epsilon_\zeta(x_\zeta)\epsilon_\epsilon(x_\epsilon)\epsilon_\delta(x_\delta)\epsilon_\gamma(x_\gamma)\epsilon_\beta(x_\beta)$
$\underline{\epsilon_\alpha(x_\alpha)\epsilon_\alpha(y_\alpha)}\epsilon_\beta(y_\beta)\epsilon_\gamma(y_\gamma)\epsilon_\delta(y_\delta)\epsilon_\epsilon(y_\epsilon)\epsilon_\zeta(y_\zeta) =$

$\epsilon_\alpha(x_\alpha + y_\alpha)\epsilon_\zeta(x_\zeta)\epsilon_\epsilon(x_\epsilon)(\epsilon_\delta(x_\delta)\epsilon_\epsilon(h(x_\alpha + y_\alpha)x_\delta)\epsilon_\gamma(x_\gamma)\epsilon_\delta(e(x_\alpha + y_\alpha)x_\gamma)$
$\epsilon_\epsilon(f(x_\alpha + y_\alpha)^2 x_\gamma)\epsilon_\zeta(g(x_\alpha + y_\alpha)x_\gamma^2))(\underline{\epsilon_\beta(x_\beta)}\epsilon_\gamma(a(x_\alpha + y_\alpha)x_\beta)\epsilon_\delta(b(x_\alpha + y_\alpha)^2 x_\beta)$
$\epsilon_\epsilon(c(x_\alpha + y_\alpha)^3 x_\beta)\epsilon_\zeta(d(x_\alpha + y_\alpha)^3 x_\beta^2)) \cdot \underline{\epsilon_\beta(y_\beta)}\epsilon_\gamma(y_\gamma)\epsilon_\delta(y_\delta)\epsilon_\epsilon(y_\epsilon)\epsilon_\zeta(y_\zeta) =$

$$\begin{aligned}
&\epsilon_\alpha(z_\alpha)\epsilon_\beta(x_\beta+y_\beta)\epsilon_\zeta(x_\zeta)(\epsilon_\epsilon(x_\epsilon)\epsilon_\zeta(i(x_\beta+y_\beta)x_\epsilon))\epsilon_\delta(x_\delta)(\epsilon_\epsilon(h(x_\alpha+y_\alpha)x_\delta)\\
&\epsilon_\zeta(ih(x_\beta+y_\beta)(x_\alpha+y_\alpha)x_\delta))\epsilon_\gamma(x_\gamma)\epsilon_\delta(e(x_\alpha+y_\alpha)x_\gamma)(\epsilon_\epsilon(f(x_\alpha+y_\alpha)^2x_\gamma)\\
&\epsilon_\zeta(fi(x_\alpha+y_\alpha)^2(x_\beta+y_\beta)x_\gamma))\epsilon_\zeta(g(x_\alpha+y_\alpha)x_\gamma^2)\cdot\\
&\epsilon_\gamma(a(x_\alpha+y_\alpha)x_\beta)\epsilon_\delta(b(x_\alpha+y_\alpha)^2x_\beta)\\
&(\epsilon_\epsilon(c(x_\alpha+y_\alpha)^3x_\beta)\epsilon_\zeta(ci(x_\alpha+y_\alpha)^3x_\beta y_\beta))\epsilon_\zeta(d(x_\alpha+y_\alpha)^3x_\beta^2)\cdot\epsilon_\gamma(y_\gamma)\epsilon_\delta(y_\delta)\\
&\epsilon_\epsilon(y_\epsilon)\epsilon_\zeta(y_\zeta)=
\end{aligned}$$

$$\begin{aligned}
&\epsilon_\alpha(z_\alpha)\epsilon_\beta(z_\beta)\epsilon_\gamma(x_\gamma+y_\gamma+a(x_\alpha+y_\alpha)x_\beta)\epsilon_\zeta(x_\zeta)\epsilon_\epsilon(x_\epsilon)\epsilon_\zeta(i(x_\beta+y_\beta)x_\epsilon)\\
&(\epsilon_\delta(x_\delta)\epsilon_\zeta(j(x_\gamma+y_\gamma)x_\delta+aj(x_\alpha+y_\alpha)x_\beta x_\delta))\\
&\epsilon_\epsilon(h(x_\alpha+y_\alpha)x_\delta)\epsilon_\zeta(ih(x_\beta+y_\beta)(x_\alpha+y_\alpha)x_\delta)\cdot\\
&\epsilon_\delta(e(x_\alpha+y_\alpha)x_\gamma)\epsilon_\zeta(ej(x_\alpha+y_\alpha)y_\gamma x_\gamma+aje(x_\alpha+y_\alpha)^2x_\beta x_\gamma))\epsilon_\epsilon(f(x_\alpha+y_\alpha)^2x_\gamma)\\
&\epsilon_\zeta(fi(x_\alpha+y_\alpha)^2(x_\beta+y_\beta)x_\gamma)\epsilon_\zeta(g(x_\alpha+y_\alpha)x_\gamma^2)\cdot(\epsilon_\delta(b(x_\alpha+y_\alpha)^2x_\beta)\\
&\epsilon_\zeta(bj(x_\alpha+y_\alpha)^2x_\beta y_\gamma))\epsilon_\epsilon(c(x_\alpha+y_\alpha)^3x_\beta)\epsilon_\zeta(ci(x_\alpha+y_\alpha)^3x_\beta y_\beta)\\
&\epsilon_\zeta(d(x_\alpha+y_\alpha)^3x_\beta^2)\cdot\epsilon_\delta(y_\delta)\epsilon_\epsilon(y_\epsilon)\epsilon_\zeta(y_\zeta)=
\end{aligned}$$

$$\begin{aligned}
&\epsilon_\alpha(z_\alpha)\epsilon_\beta(z_\beta)\epsilon_\gamma(z_\gamma)\epsilon_\delta(x_\delta+y_\delta+b(x_\alpha+y_\alpha)^2x_\beta+e(x_\alpha+y_\alpha)x_\gamma)\epsilon_\zeta(x_\zeta)\epsilon_\epsilon(x_\epsilon)\\
&\epsilon_\zeta(i(x_\beta+y_\beta)x_\epsilon)\cdot\epsilon_\zeta(j(x_\gamma+y_\gamma)x_\delta+aj(x_\alpha+y_\alpha)x_\beta x_\delta)\epsilon_\epsilon(h(x_\alpha+y_\alpha)x_\delta)\\
&\epsilon_\zeta(ih(x_\beta+y_\beta)(x_\alpha+y_\alpha)x_\delta)\\
&\epsilon_\zeta(ej(x_\alpha+y_\alpha)y_\gamma x_\gamma+aje(x_\alpha+y_\alpha)^2x_\beta x_\gamma)\epsilon_\epsilon(f(x_\alpha+y_\alpha)^2x_\gamma)\\
&\epsilon_\zeta(fi(x_\alpha+y_\alpha)^2(x_\beta+y_\beta)x_\gamma)\epsilon_\zeta(g(x_\alpha+y_\alpha)x_\gamma^2)\\
&\epsilon_\zeta(bj(x_\alpha+y_\alpha)^2x_\beta y_\gamma)\epsilon_\epsilon(c(x_\alpha+y_\alpha)^3x_\beta)\\
&\epsilon_\zeta(ci(x_\alpha+y_\alpha)^3x_\beta y_\beta)\epsilon_\zeta(d(x_\alpha+y_\alpha)^3x_\beta^2)\cdot\epsilon_\epsilon(y_\epsilon)\epsilon_\zeta(y_\zeta)=
\end{aligned}$$

$$\begin{aligned}
&\epsilon_\alpha(z_\alpha)\epsilon_\beta(z_\beta)\epsilon_\gamma(z_\gamma)\epsilon_\delta(z_\delta)\\
&\epsilon_\epsilon(x_\epsilon+y_\epsilon+c(x_\alpha+y_\alpha)^3x_\beta+f(x_\alpha+y_\alpha)^2x_\gamma+h(x_\alpha+y_\alpha)x_\delta)\\
&\epsilon_\zeta(x_\zeta)\cdot\epsilon_\zeta(i(x_\beta+y_\beta)x_\epsilon)\epsilon_\zeta(j(x_\gamma+y_\gamma)x_\delta+aj(x_\alpha+y_\alpha)x_\beta x_\delta)\\
&\epsilon_\zeta(ih(x_\beta+y_\beta)(x_\alpha+y_\alpha)x_\delta)\\
&\epsilon_\zeta(ej(x_\alpha+y_\alpha)y_\gamma x_\gamma+aje(x_\alpha+y_\alpha)^2x_\beta x_\gamma)\\
&\epsilon_\zeta(fi(x_\alpha+y_\alpha)^2(x_\beta+y_\beta)x_\gamma)\epsilon_\zeta(g(x_\alpha+y_\alpha)x_\gamma^2)\\
&\epsilon_\zeta(bj(x_\alpha+y_\alpha)^2x_\beta y_\gamma)\cdot\epsilon_\zeta(ci(x_\alpha+y_\alpha)^3x_\beta y_\beta)\epsilon_\zeta(d(x_\alpha+y_\alpha)^3x_\beta^2)\cdot\epsilon_\zeta(y_\zeta)=
\end{aligned}$$

$$\begin{aligned}
&\epsilon_\alpha(z_\alpha)\epsilon_\beta(z_\beta)\epsilon_\gamma(z_\gamma)\epsilon_\delta(z_\delta)\epsilon_\epsilon(z_\epsilon)\\
&\epsilon_\zeta(x_\zeta+i(x_\beta+y_\beta)x_\epsilon+j(x_\gamma+y_\gamma)x_\delta+aj(x_\alpha+y_\alpha)x_\beta x_\delta+ih(x_\beta+y_\beta)(x_\alpha+y_\alpha)x_\delta)\\
&\epsilon_\zeta(ej(x_\alpha+y_\alpha)y_\gamma x_\gamma+aje(x_\alpha+y_\alpha)^2x_\beta x_\gamma+\\
&\qquad\qquad fi(x_\alpha+y_\alpha)^2(x_\beta+y_\beta)x_\gamma+g(x_\alpha+y_\alpha)x_\gamma^2)\\
&\epsilon_\zeta(bj(x_\alpha+y_\alpha)^2x_\beta y_\gamma+ci(x_\alpha+y_\alpha)^3x_\beta y_\beta+d(x_\alpha+y_\alpha)^3x_\beta^2+y_\zeta).
\end{aligned}$$

COROLLARY 2.2. *$t^{-1}n^{-1}tn =$*

$$\epsilon_\alpha(x(\alpha))\epsilon_\beta(x(\beta))\epsilon_\gamma(x(\gamma))\epsilon_\delta(x(\delta))\epsilon_\epsilon(x(\epsilon))\epsilon_\zeta(x(\zeta))$$

where

$$\begin{aligned}
&n=\epsilon_\alpha(n_\alpha)\epsilon_\alpha(n_\alpha)\epsilon_\beta(n_\beta)\epsilon_\gamma(n_\gamma)\epsilon_\delta(n_\delta)\epsilon_\zeta(n_\zeta)\\
&x(\alpha)=(1-\alpha^{-1})n_\alpha\\
&x(\beta)=(1-\beta^{-1})n_\beta\\
&x(\gamma)=(1-\gamma^{-1})n_\gamma-a(1-\alpha^{-1})\beta^{-1}n_\alpha n_\beta\\
&x(\delta)=(1-\delta^{-1})n_\delta-b(1-\alpha^{-1})^2\beta^{-1}n_\alpha^2n_\beta-e(1-\alpha^{-1})\gamma^{-1}n_\alpha n_\gamma
\end{aligned}$$

$$
\begin{aligned}
&x(\epsilon) = (1-\epsilon^{-1})n_\epsilon - c(1-\alpha^{-1})^3\beta^{-1}n_\alpha^3 n_\beta - f(1-\alpha^{-1})^2\gamma^{-1}n_\alpha n_\gamma \\
&\quad -h(1-\alpha^{-1})\delta^{-1}n_\alpha n_\delta \\
&x(\zeta) = (1-\zeta^{-1})n_\zeta + d(1-\alpha^{-1})^3\beta^{-2}n_\alpha^3 n_\beta^2 + g(1-\alpha^{-1})\gamma^{-2}n_\alpha n_\gamma^2 \\
&\quad -i(1-\beta^{-1})\epsilon^{-1}n_\beta n_\epsilon - j(1-\gamma^{-1})\delta^{-1}n_\gamma n_\delta \\
&\quad -hi(1-\alpha^{-1})(1-\beta^{-1})\delta^{-1}n_\alpha n_\beta n_\delta \\
&\quad -fi(1-\alpha^{-1})^2(1-\beta^{-1})\gamma^{-1}n_\alpha^2 n_\beta n_\gamma - ci(1-\alpha^{-1})^3\beta^{-1}n_\alpha^3 n_\beta^2 \\
&\quad -bj(1-\alpha^{-1})^2\beta^{-1}n_\alpha^2 n_\beta n_\gamma - ej(1-\alpha^{-1})\gamma^{-1}n_\alpha n_\gamma^2 \\
&\quad +aj(1-\alpha^{-1})\beta^{-1}\delta^{-1}n_\alpha n_\beta n_\delta + aje(1-\alpha^{-1})^2\beta^{-1}\gamma^{-1}n_\alpha^2 n_\beta n_\gamma.
\end{aligned}
$$

PROOF. Let $x_\eta = -n_\eta \eta^{-1}, y_\eta = n_\eta$ in the previous lemma for $\eta = \alpha, \beta, \gamma, \delta, \epsilon, \zeta$.

In the next lemma we gather together equations that will be used to prove the main result of this section.

LEMMA 2.3. *The following equations hold on Y''.*

(a) $w(\alpha) = 1, w(\beta) = 1$

(b) $n_\alpha = 1/z(W_+,\alpha), n_\beta = 1/z(W_+,\beta)$,

(c) $\lambda = x(\alpha)z(\alpha), \lambda = x(\beta)z(\beta)$

(d) $(1-\alpha^{-1}) = x(\alpha)z(W_+,\alpha), (1-\beta^{-1}) = x(\beta)z(W_+,\beta)$

(e) $z(W_+,\alpha)/z(\alpha) = z_1(W_+,\alpha) = (1-\alpha^{-1})/\lambda$
$z(W_+,\beta)/z(\beta) = z_1(W_+,\beta) = (1-\beta^{-1})/\lambda$

(f) $u_2' n_\gamma = -1/(z(W_+,\beta)z(W(\sigma_\beta),\alpha))$ *where* $\sigma_\beta(X_\gamma) = u_2' X_\alpha$

(g)

$$
\begin{aligned}
\lambda w(\gamma) &= z(\alpha)z(\beta)x(\gamma) \\
&= z(\alpha)z(\beta)((1-\gamma^{-1})n_\gamma - \beta^{-1}(1-\alpha^{-1})n_\alpha n_\beta) \\
&= [\lambda^2/(1-\alpha^{-1})(1-\beta^{-1})] \quad \textit{times} \\
&\qquad [(-(1-\gamma^{-1})u_2'^{-1}z(W_+,\alpha)/z(W(\sigma_\beta),\alpha)) - \beta^{-1}(1-\alpha^{-1})]
\end{aligned}
$$

PROOF. (a) holds by definition. (b) was proved in (I.5.5). (c) are the defining relations for $z(\alpha)$ and $z(\beta)$. (d) is equation (II.3.3) combined with (b). In (e) the first equality on each line serves as the definition of $z_1(W,-)$. The second equality on each line is obtained by dividing (d) by (c). (f) was proved for G_2 in (II.7.2), (II.7.3). This calculation only makes use of the fact that β is at least as long as α, so that the calculation holds for A_2 and B_2 as well. This is consistent with (II.8.1) when $\alpha = \beta_2, \beta = \beta_1, u_2' = -1$. The first equality of (g) is (II.4.2). The second equality is (2.2). This is consistent with (II.5.1). The third equality of (g) follows by using (e) for $z(\alpha)$ and $z(\beta)$, (f) for n_γ, and (b) for $n_\alpha n_\beta$.

LEMMA 2.4. *Let $dx_1 \dots dx_n$ be a form of maximal degree on Y_Γ. Suppose that there are coordinates $\mu_1, \dots, \mu_n$ on Y_Γ such that locally $\mu_1 = 0$ defines a divisor E and $x_i = \mu_1^{a_i}\xi_i$ where ξ_i is regular on E. Let $a = \sum a_i$. Then $dx_1 \dots dx_n$ vanishes to order at least $a-1$ on E.*

PROOF.

$$dx_1 \dots dx_n = \frac{\partial(x_1,\dots,x_n)}{\partial(\mu_1,\dots,\mu_n)}\, d\mu_1 \dots d\mu_n.$$

Expanding the Jacobian by the first row we obtain

$$\frac{\partial(x_1,\dots,x_n)}{\partial(\mu_1,\dots,\mu_n)} = \sum(\pm 1)\,\frac{\partial x_i}{\partial \mu_1}\,\frac{\partial(x_1,\dots,\hat{x}_i,\dots,x_n)}{\partial(\mu_2,\dots,\mu_n)}.$$

Now

$$\frac{\partial(x_1,\dots,\hat{x}_i,\dots,x_n)}{\partial(\mu_2,\mu_3,\dots,\mu_n)} = \mu_1^{a-a_i}\,\frac{\partial(\xi_1,\dots,\hat{\xi}_i,\dots,\xi_n)}{\partial(\mu_2,\mu_3,\dots,\mu_n)}$$

and

$$\frac{\partial x_i}{\partial \mu_1} = a_i\mu_1^{a_i-1}\xi_i + \mu_1^{a_i}\frac{\partial \xi_i}{\partial \mu_1} = \mu_1^{a_i-1}(a_i\xi_i + \mu_1\frac{\partial \xi_i}{\partial \mu_1}).$$

Thus every term of the sum vanishes to order at least $(a-a_i)+(a_i-1) = a-1$.

3. Exclusion of Spurious Divisors.

The next few lemmas show that divisors with certain zero patterns make no contribution to the subregular germ. A subregular unipotent element is one whose centralizer has dimension $rank(G)+2$. The proofs follow similar lines. Let E be a divisor. If λ vanishes to order a on E and the form ω_Y vanishes to order $b-1$ on E then E makes a contribution to a term $m(\lambda)^r\theta(\lambda)|\lambda|^{\beta-1}F_r(\theta,\beta,f)$ of the asymptotic expansion with $\beta = b/a$. For details see [15].

Suppose we can express ω_Y locally as $\lambda^2\mu^x(d\mu/\mu)\wedge\omega'$ where $x>0,\mu$ is a local coordinate, $\mu=0$ defines E, and ω' regular on E. Then $b=2a+x, \beta-1 = (b/a)-1 = 1+(x/a) > 1$. This shows that such a divisor E does not contribute to the first order term of the asymptotic expansion. By (*II*.9.2), it does not contribute to the subregular germ.

The coordinate functions $z(W,\alpha)$ are regular on $Y_1(B_\infty,B_0)$. However, it is rather awkward to work directly with these coordinates. On the variety Y'' we have seen in chapter II how to express the functions $w(\gamma)$ in terms of $z(\alpha)$ (α simple) and the coefficients of t and n (*II*.3) and also how to express the coefficients of n in terms of the coordinates $z(W,\alpha)$. Also $z(\alpha) = (\lambda/(1-\alpha^{-1}))z(W_+,\alpha)$ (2.3). Thus on Y'' we can express $w(\eta)$ in terms of the coefficients of t,λ and $z(W,\alpha)$. We use this expression to extend $w(\gamma)$ to a rational function on $Y_1(B_\infty,B_0)$. Similarly we extend n_γ to a rational function on $Y_1(B_\infty,B_0)$. The following lemmas show that with appropriate hypotheses, $w(\eta)$ or sometimes $1/w(\eta)$ may actually extend to a regular function on a Zariski open set of certain spurious divisors.

The next lemma does not assume that G is rank two.

LEMMA 3.1. *Let E be a divisor on Y_Γ. Suppose that W_+ and two simple adjacent roots α_1 and α_2 can be chosen so that $\theta_1(W_+,\alpha_1)\neq 0$, and $\theta_1(W(\sigma_{\alpha_2}),\alpha_1) = 0$ on E. Suppose also that the root α_1 is not longer than the root α_2. Then*

a) $1/w(\alpha_1+\alpha_2)$ is regular on E.

b) $1/w(\alpha_1+\alpha_2)$ vanishes on E.

c) E makes no contribution to the subregular germ.

PROOF. Note that $z(\alpha) = \lambda z(W_+,\alpha)/(1-\alpha^{-1})$ and $z(\beta) = \lambda z(W_+,\beta)/(1-\beta^{-1})$ are regular on E. Set $w = 1/w(\alpha_1+\alpha_2)$. On Y'' we have by (2.3.g)

$w = [\lambda^2/(1-\alpha_1^{-1})(1-\alpha_2^{-1})]^{-1}$

$$[(-(1-(\alpha_1\alpha_2)^{-1})/\lambda)(u_2'^{-1}z(W_+,\alpha_1)/z(W(\sigma_{\alpha_2}),\alpha_1)) - \alpha_2^{-1}(1-\alpha_1^{-1})/\lambda]^{-1}.$$

Set $z = z(W(\sigma_{\alpha_2}),\alpha_1)/z(W_+,\alpha_1)$. By the hypotheses of the lemma, z is regular on E and is in fact equal to zero on E. Thus up to a regular invertible factor, w is equal to

$$z[-((1-(\alpha_1\alpha_2)^{-1})/\lambda)u_2'^{-1} - z\alpha_2^{-1}(1-\alpha_1^{-1})/\lambda]^{-1}.$$

Since $z = 0$ on E, the factor

$$-((1-(\alpha_1\alpha_2^{-1}))/\lambda)u_2'^{-1} - z\alpha_2^{-1}((1-\alpha_1^{-1})/\lambda)$$

equals $-((1-(\alpha_1\alpha_2)^{-1})/\lambda)u_2'^{-1}$ on E and is consequently regular and invertible on E. Parts a) and b) follow.

Formulas (2.3.c,g) give

$$x(\alpha_1) = \lambda/z(\alpha_1) = z(\alpha_2)x(\alpha_1+\alpha_2)w, \text{ and}$$
$$x(\alpha_2) = \lambda/z(\alpha_2) = z(\alpha_1)x(\alpha_1+\alpha_2)w,$$
$$\lambda = z(\alpha_1)z(\alpha_2)x(\alpha_1+\alpha_2)w.$$

The form up to a factor we can ignore is given on Y^0 by
$d\lambda \wedge dx(\alpha_1) \wedge dx(\alpha_2) \wedge dx(\alpha_1+\alpha_2) \wedge \ldots d\nu =$
$\lambda^2 d\lambda \wedge d(1/z(\alpha_1)) \wedge d(1/z(\alpha_2)) \wedge dx(\alpha_1+\alpha_2) \wedge \ldots d\nu =$
$\lambda^2 x(\alpha_1+\alpha_2)dw \wedge (dz(\alpha_1)/z(\alpha_1)) \wedge (dz(\alpha_2)/z(\alpha_2)) \wedge dx(\alpha_1+\alpha_2) \wedge \ldots d\nu$
where

$$d\nu = d\nu_1 \wedge d\nu_2 \wedge \ldots \wedge d\nu_p.$$

Let $\mu = \mu_1, \ldots, \mu_n$ be a coordinate system on Y_Γ near a point of E, and suppose that $\mu = 0$ defines the divisor E locally. Now pull the variables $\lambda, x(\alpha_1+\alpha_2), w, z(\alpha_1), z(\alpha_2)$ etc. up to Y_Γ and write

$$z(\alpha_1) = \mu^{e_1}\xi_1$$
$$z(\alpha_2) = \mu^{e_2}\xi_2$$
$$x(\alpha_1+\alpha_2) = \mu^{e_3}\xi_3$$
$$w = \mu^{e_4}\xi_4$$

where ξ_i is regular on E. By (2.4), $dw \wedge dz(\alpha_1) \wedge dz(\alpha_2) \wedge dx(\alpha_1+\alpha_2)$ vanishes to order at least $e_4+e_1+e_2+e_3-1$. Thus

$$\omega_Y = \lambda^2\mu^x(d\mu/\mu) \wedge \omega'$$

where

$$x \geq (e_3-e_1-e_2)+(e_4+e_1+e_2+e_3-1)+1 = 2e_3+e_4,$$

and ω' is regular on E. But w vanishes on E so $e_4 > 0$. By remarks at the beginning of this section, the proof is complete.

LEMMA 3.2. *(B_2) Suppose that the Weyl chamber W_+ can be chosen on E so that $\theta_1(W_+,\alpha) \neq 0, \theta_1(W_+,\beta) \neq 0, \theta_1(W(\sigma_\beta),\alpha) \neq 0, \theta_1(W(\sigma_\alpha),\beta) = 0$. Then*
a) $w(\gamma)$ is regular on E,
b) $1/w(\delta)$ is regular on E and in fact vanishes on E,
c) E makes no contribution to the subregular germ.

REMARK. This lemma treats the zero pattern $B.IV$.

PROOF. By (2.3.e) $z(\alpha) = (\lambda/(1-\alpha^{-1}))z(W_+,\alpha)$ and

$$z(\beta) = (\lambda/(1-\beta^{-1}))z(W_+,\beta).$$

Also by (2.3.b,f)

$$n_\gamma/(n_\alpha n_\beta) = -u_2'^{-1} z(W_+,\alpha)/z(W(\sigma_\beta),\alpha).$$

Thus by the assumptions of the lemma $z(\alpha), z(\beta)$, and $n_\gamma/(n_\alpha n_\beta)$ are regular. By (2.3.g)

$$\begin{aligned} w(\gamma) &= z(\alpha)z(\beta)((1-\gamma^{-1})/\lambda)n_\gamma - \beta^{-1}((1-\alpha^{-1})/\lambda)n_\alpha n_\beta) = \\ &= (\lambda^2/(1-\alpha^{-1})(1-\beta^{-1}))(((1-\gamma^{-1})/\lambda)(n_\gamma/(n_\alpha n_\beta)) \\ &\quad - \beta^{-1}((1-\alpha^{-1})/\lambda)). \end{aligned}$$

Thus the regularity of $n_\gamma/(n_\alpha n_\beta)$ implies the regularity of $w(\gamma)$. This proves (a).

Let $z = z(W(\sigma_\alpha),\beta)/z(W_+,\beta)$. Then $z = 0$ on E by assumption. By (II.3.3),

$$\begin{aligned} zw(\delta) &= z(1-\delta^{-1})n_\delta z(\alpha)^2 z(\beta)/\lambda + zz(\alpha)^2 z(\beta) \sum c_{\beta_1\ldots\beta_n}(t) n_{\beta_1}\ldots n_{\beta_n}/\lambda \\ z(\alpha) &= \lambda/((1-\alpha^{-1})n_\alpha), z(\beta) = \lambda/((1-\beta^{-1})n_\beta). \end{aligned}$$

Thus

$$zw(\delta) = [\lambda^3/((1-\alpha^{-1})^2(1-\beta^{-1}))][zq + (z(1-\delta^{-1})n_\delta/(\lambda n_\alpha^2 n_\beta))]$$

where q is the regular function $q = c_{\alpha\alpha\beta}(t)/\lambda + c_{\alpha\gamma}(t)n_\gamma/(\lambda n_\alpha n_\beta)$. Since q is regular on E and $z = 0$ on $E, zq = 0$ on E. If $zw(\delta)$ is regular and invertible on E then (b) will follow. $\lambda^2(1-\delta^{-1})/((1-\alpha^{-1})^2(1-\beta^{-1}))$ is regular and invertible on E and zq vanishes on E, so $zw(\delta)$ is regular and invertible on E if and only if $zn_\delta/(n_\alpha^2 n_\beta)$ is regular and invertible on E. The following lemma completes the proof of (b).

LEMMA 3.3. *(B_2) Let E be as in lemma 3.2. Then $zn_\delta/(n_\alpha^2 n_\beta)$ equals a non-zero constant on E.*

PROOF. We apply the condition $B\omega n = Bn_w$ to the situation $\omega = \sigma_\beta\sigma_\alpha$ and

$$n_w = \exp(z_2 X_{-\beta})\exp(z_1 X_{-\alpha})$$
$$\text{with } z_2 = z(W(\sigma_\alpha),\beta), z_1 = z(W_+,\alpha).$$

This condition can be rewritten

$$\sigma_\beta\sigma_\alpha\epsilon_\alpha(n_\alpha)\epsilon_\beta(n_\beta)\epsilon_\gamma(n_\gamma)\epsilon_\delta(n_\delta)\epsilon_{-\alpha}(-z_1)\epsilon_{-\beta}(-z_2)\in B$$

or

$$\sigma_\beta\{\sigma_\alpha\epsilon_\alpha(n_\alpha)\epsilon_{-\alpha}(-z_1)\}\{\epsilon_{-\alpha}(z_1)\epsilon_\beta(n_\beta)\epsilon_\gamma(n_\gamma)\epsilon_\delta(n_\delta)\epsilon_{-\alpha}(-z_1)\}\epsilon_{-\beta}(-z_2)\in B.$$

By (I.5.5) the first bracketed term equals $(-z(W_+,\alpha))^{\alpha^v}\epsilon_\alpha(-n_\alpha)$ because $z_1 n_\alpha = 1$. Thus

$$\sigma_\beta\{\sigma_\alpha\epsilon_\alpha(n_\alpha)\epsilon_{-\alpha}(-z_1)\}\sigma_\beta^{-1}$$

lies in B. The condition becomes

$$\sigma_\beta\{\epsilon_{-\alpha}(z_1)\epsilon_\beta(n_\beta)\epsilon_\gamma(n_\gamma)\epsilon_\delta(n_\delta)\epsilon_{-\alpha}(-z_1)\}\epsilon_{-\beta}(-z_2)\in B.$$

Now we have

$$\begin{aligned}\epsilon_{-\alpha}(z_1)\epsilon_\beta(n_\beta)\epsilon_{-\alpha}(-z_1) &= \epsilon_\beta(n_\beta)\\ \epsilon_{-\alpha}(z_1)\epsilon_\gamma(n_\gamma)\epsilon_{-\alpha}(-z_1) &= \epsilon_\beta(ez_1n_\gamma) \text{ modulo } N_\beta\\ \epsilon_{-\alpha}(z_1)\epsilon_\delta(n_\delta)\epsilon_{-\alpha}(-z_1) &= \epsilon_\beta(fz_1^2n_\delta) \text{ modulo } N_\beta\end{aligned}$$

for some non-zero constants e and f. Thus

$$\epsilon_{-\alpha}(z_1)\epsilon_\beta(n_\beta)\epsilon_\gamma(n_\gamma)\epsilon_\delta(n_\delta)\epsilon_{-\alpha}(-z_1) = \epsilon_\beta(y) \text{ modulo } N_\beta,$$

$$y = n_\beta + ez_1n_\gamma + fz_1^2n_\delta.$$

Thus by (e.g. II.6.1), $1 - z_2 y = 0$, or

$$\begin{aligned}1/z_2 &= n_\beta + ez_1n_\gamma + fz_1^2n_\delta, z_1 = 1/n_\alpha\\ 1/(z_2n_\beta) &= 1 + en_\gamma/(n_\alpha n_\beta) + fn_\delta/(n_\alpha^2 n_\beta)\\ fzn_\delta/(n_\alpha^2n_\beta) &= -z - zen_\gamma/(n_\alpha n_\beta) + z/(z_2n_\beta).\end{aligned}$$

The first two terms on the right hand side of this last equation vanish on E because $n_\gamma/(n_\alpha n_\beta)$ is regular on E and z vanishes on E. Also

$$z/(z_2n_\beta) = (z(W(\sigma_\alpha),\beta)/z(W_+,\beta))(z(W_+,\beta)/z(W(\sigma_\alpha),\beta)) = 1.$$

This completes the proof of lemma 3.3.

We now continue with the proof of lemma 3.2.c. We use coordinates $w(\gamma)$ and $w' = 1/w(\delta)$. The relation *II.4.2* gives $w(\delta)x(\gamma) = w(\gamma)x(\delta)z(\alpha)$ and $\lambda w(\delta) = x(\delta)z(\alpha)^2z(\beta)$ or

$$x(\gamma) = w'w(\gamma)x(\delta)z(\alpha) \text{ and } \lambda = w'x(\delta)z(\alpha)^2z(\beta).$$

The form up to a factor we can ignore is given on Y^0 by

$$d\lambda \wedge dx(\alpha) \wedge dx(\beta) \wedge dx(\gamma) \wedge dx(\delta) \wedge \ldots d\nu =$$

$$\lambda^2 d\lambda \wedge d(1/z(\alpha)) \wedge d(1/z(\beta)) \wedge d(w'w(\gamma)x(\delta)z(\alpha)) \wedge dx(\delta) \wedge \ldots d\nu =$$

$$\lambda^2 x(\delta)^2 w' z(\alpha)^2 dw' \wedge (dz(\alpha)/z(\alpha)) \wedge (dz(\beta)/z(\beta)) \wedge dw(\gamma) \wedge dx(\delta) \wedge \ldots d\nu$$

where

$$d\nu = d\nu_1 \wedge d\nu_2 \wedge \ldots \wedge d\nu_p.$$

Let $\mu = \mu_1, \ldots, \mu_n$ be a coordinate system on Y_Γ near a point of E, and suppose that $\mu = 0$ defines the divisor E locally. As before pull the variables $\lambda, z(\alpha), z(\beta), w(\gamma), w'$, etc., up to Y_Γ and write

$$\begin{aligned} z(\alpha) &= \mu^{e_1}\xi_1 \\ z(\beta) &= \mu^{e_2}\xi_2 \\ x(\delta) &= \mu^{e_3}\xi_3 \\ w(\gamma) &= \mu^{e_4}\xi_4 \\ w' &= \mu^{e_5}\xi_5. \end{aligned}$$

Then by lemma 2.4, $\omega_Y = \lambda^2\mu^x(d\mu/\mu)\wedge\omega'$, $x \geq (2e_3+e_5+2e_1)+(e_5+e_4+e_3)$. But $w' = 0$ on E so that $e_5 > 0$. The remarks at the beginning of the section now give the result.

REMARK 3.4. Proposition 1.1 has now been proved for special nodes of type A_2 and B_2. This follows by an examination of the zero patterns for these groups. Lemma 3.1 covers the A_2 and all the zero patterns of B_2 except for the pattern $B.IV$. Lemma 3.2 treats the pattern $B.IV$.

LEMMA 3.5. *(G_2) Suppose that W_+ can be selected so that on E*

$$\begin{aligned} \theta_1(W_+, \alpha) &\neq 0, \\ \theta_1(W_+, \beta) &\neq 0, \\ \theta_1(W(\sigma_\beta), \alpha) &\neq 0, \end{aligned}$$

$$(z_1(W(\sigma_\beta\sigma_\alpha), \alpha)z_1(W(\sigma_\alpha), \beta))/(z_1(W_+, \alpha)z_1(W_+, \beta)) = 0,$$

and

$$z(W(\sigma_\beta\sigma_\alpha), \alpha)/z(W_+, \alpha) \neq -1$$

on E then
a) $w(\gamma)$ is regular on E
b) $1/w(\delta)$ is regular on E and in fact vanishes.
c) E makes no contribution to the subregular germ.

PROOF. Let $z = z(W(\sigma_\beta\sigma_\alpha), \alpha)z(W(\sigma_\alpha), \beta)/z(W_+, \alpha)z(W_+, \beta)$. Then $z = 0$ on E. It follows from the expression (2.3.b,f) for $n_\gamma/(n_\alpha n_\beta)$ that it is regular on

E. (a) now follows by the same argument used in lemma 3.2. Also $zn_\gamma/(n_\alpha n_\beta) = 0$ on E. (b) will follow if I prove that $zw(\delta)$ is regular and invertible on E. An argument identical to that in lemma 3.2 shows that $zw(\delta)$ is regular and invertible on E if and only if $zn_\delta/(n_\alpha^2 n_\beta)$ has the same property. By lemma II.7.2, $m_\delta/(m_\alpha^2 m_\beta) = n_\delta/(n_\alpha^2 n_\beta) + b - en_\gamma/(n_\alpha n_\beta)$ where b and e are constants defined in chapter II. Since b and $en_\gamma/(n_\alpha n_\beta)$ are regular on E, $zn_\delta/(n_\alpha^2 n_\beta)$ is regular and invertible on E if and only if $zm_\delta/(m_\alpha^2 m_\beta)$ is regular and invertible on E. By lemma II.7.3, we have

$$zm_\delta/(m_\alpha^2 m_\beta) = (z(W_+,\alpha) + z(W(\sigma_\beta\sigma_\alpha),\alpha)/z(W_+,\alpha)$$

which is non-zero by hypothesis. This proves (b).

Now the expression for ω_Y contained in the proof of lemma 3.2 shows that $w' = 0$ implies (c).

LEMMA 3.6. *(G_2) Suppose that W_+ can be chosen so that $\theta_1(W_+,\alpha) \neq 0$, $\theta_1(W_+,\beta) \neq 0, \theta_1(W(\sigma_\beta),\ \alpha) \neq 0, \theta_1(W(\sigma_\beta\sigma_\alpha),\ \alpha) \neq 0, \theta_1(W(\sigma_\alpha),\beta) \neq 0$, and $\theta_1(W(\sigma_\alpha\sigma_\beta),\beta) = 0$ on E. Then*
a) $w(\gamma), w(\delta)$, and $w(\epsilon)$ are regular on E
b) $w'' = 1/w(\zeta)$ is regular and vanishes on E.
c) E is not a subregular divisor.

PROOF. That $z(\alpha), z(\beta), n_\gamma/(n_\alpha n_\beta), n_\delta/(n_\alpha^2 n_\beta), n_\epsilon/(n_\alpha^3 n_\beta)$ are regular on E follows from lemmas II.7.2 and II.7.3. That $w(\gamma), w(\delta)$, and $w(\epsilon)$ are regular on E now follows directly from lemma II.3.3.

Set $z = z(W(\sigma_\alpha\sigma_\beta),\beta)/z(W_+,\beta)$. To prove (b) we show that $zw(\zeta)$ is regular and invertible on E. Proceeding as in the previous lemma we see that $zw(\zeta)$ is regular and invertible on E if and only if $zm_\zeta/(m_\alpha^3 m_\beta)$ is regular and invertible on E. But by (II.7.3),

$$m_\zeta/(m_\alpha^3 m_\beta) = (z(W_+,\alpha)^3/z(W(\sigma_\beta),\alpha)^3)(-z(W_+,\beta)/z(W(\sigma_\alpha\sigma_\beta),\beta) - 1).$$

$$\text{So } zm_\zeta/(m_\alpha^3 m_\beta) = -z(W_+,\alpha)^3/z(W(\sigma_\beta),\alpha)^3 \text{ on } E.$$

By assumption this is regular and invertible on E.

To prove (c) we note that the following equations hold (II.4.2)

$$w(\zeta)x(\eta) = w(\eta)x(\zeta)\prod z(\alpha)^{m(\alpha)} \text{ with } \zeta - \eta = \sum m(\alpha)\alpha$$

$$\text{or } x(\eta) = w''w(\eta)x(\zeta)\prod z(\alpha)^{m(\alpha)} = 0 \text{ on } E \text{ when } \eta \neq \zeta.$$

So E is certainly not a subregular divisor.

LEMMA 3.7. *(G_2) Suppose that W_+ cannot be chosen to satisfy any of the previous cases and at least one wall vanishes. Then the zero pattern must be G.I.*

PROOF. By the hypothesis of (3.1), we may assume that all of the walls of type α are non-zero (i.e., $\theta_1(W,\alpha) \neq 0 \;\forall\; W$). If the walls of type α are non-zero and the hypotheses of (3.6) fail, then there cannot be two consecutive walls of type β which are non-zero (e.g., (W_+,β) and $(W(\sigma_\alpha),\beta)$). Choose W_+ so that $\theta_1(W_+,\beta) \neq 0$. (Thus $\theta_1(W(\sigma_\alpha),\beta) = \theta_1(W(\sigma_\alpha\sigma_\beta),\beta) = 0$.) By the hypothesis of (3.5), $z_1(W(\sigma_\beta\sigma_\alpha),\alpha)/z_1(W_+,\alpha) = -1$ and by symmetry $z_1(W(\sigma_\beta),\alpha)/z_1(W(\sigma_\beta\sigma_\alpha\sigma_\beta),\alpha) = -1$. We have the following situation:

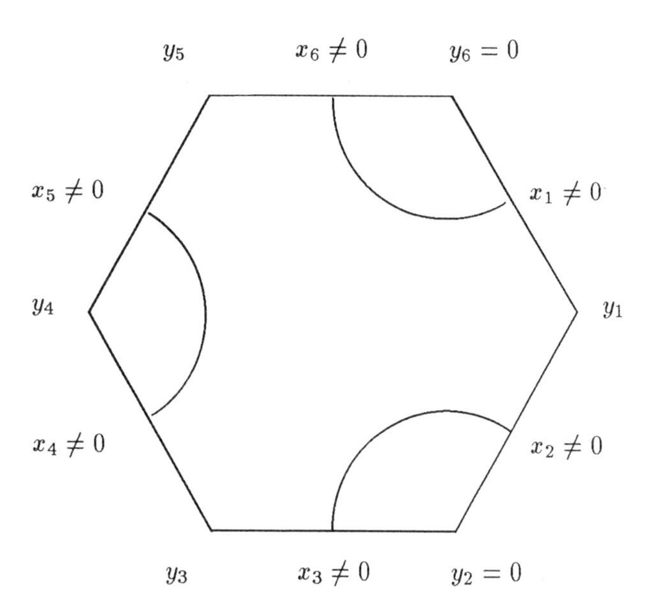

$x_2 + x_3 = 0; x_1 + x_6 = 0$ (so also $x_4 + x_5 = 0$). Now

$$\epsilon_\alpha(x_1)\epsilon_\beta(y_1)\ldots\epsilon_\alpha(x_6)\epsilon_\beta(y_6) = \epsilon_\alpha(x_5)\epsilon_\beta(y_5 + y_1 + y_3)\epsilon_\alpha(x_4)\epsilon_\beta(y_4) = 0.$$

By (1.3.b), using $x_4, x_5 \neq 0$ we have $y_4 = y_5 + y_1 + y_3 = 0$. This is pattern $(G.I)$. (We allow the possibility that $y_1 y_3 y_5 = 0$).

The following lemma completes the proof of proposition 1.1.

LEMMA 3.8. *(G_2) Suppose that E gives the pattern $G.I$. Select W_+ so that $\theta_1(W_+,\beta) = 0$, $\theta_1(W_+,\alpha) \neq 0$, $\theta_1(W(\sigma_\beta),\alpha) \neq 0$, $\theta_1(W(\sigma_\beta\sigma_\alpha),\alpha) \neq 0$, $\theta_1(W(\sigma_\alpha),\beta) \neq 0$, $\theta_1(W(\sigma_\alpha\sigma_\beta),\beta) \neq 0$ on E. Then*
a) $z(W_+,\alpha)/z(W(\sigma_\beta),\alpha) = 1$ *on E,*
b) $z(\beta)$ *vanishes on E,*
c) $n_\gamma/(n_\alpha n_\beta) = -1, n_\delta/(n_\alpha^2 n_\beta) = -3, n_\epsilon/(n_\alpha^3 n_\beta) = -5$, $n_\zeta/(n_\alpha^3 n_\beta^2) = 13$ *on E;*
d) $w(\eta)$ *is regular on E, $\eta = \gamma, \delta, \epsilon, \zeta$;*
e) $w(\zeta)$ *is invertible on E.*
f) *E is not a subregular divisor.*

PROOF. Notice that by the equations holding for the pattern $G.I$ we have $z_1(W(\sigma_\beta),\alpha) = z_1(W_+,\alpha)$. The minus sign has disappeared because the walls

$(W(\sigma_\beta),\alpha)$ and (W_+,α) have opposite orientations. Thus (a) and (b) follow easily. Now use lemma II.7.3,
$m_\alpha = 1/z(W_+,\alpha)$
$m_\beta = 1/z(W_+,\beta)$
$m_\gamma/(m_\alpha m_\beta) = +z(W_+,\alpha)/z(W(\sigma_\beta\sigma_\alpha),\alpha) = -1$ on E.

$$m_\delta/(m_\alpha^2 m_\beta) = \\ (z(W_+,\alpha)z(W_+,\beta))/(z(W(\sigma_\beta\sigma_\alpha),\alpha)z(W(\sigma_\alpha),\beta)) \\ + z(W_+,\beta)/z(W(\sigma_\alpha),\beta) = 0$$

on E.
$m_\epsilon/(m_\alpha^3 m_\beta) = -z(W_+,\beta)/z(W(\sigma_\alpha),\beta) = 0$ on E.
$m_\zeta/(m_\alpha^3 m_\beta^2) = -(z(W_+,\beta)z(W_+,\alpha)^3)/(z(W(\sigma_\alpha\sigma_\beta),\beta)z(W(\sigma_\beta),\alpha)^3)-$

$$z(W_+,\alpha)^3/z(W(\sigma_\beta),\alpha)^3 = -1 \text{ on } E.$$

Now apply (II.7.2). Recall that $a = b = c = d = 1, e = 2, f = g = h = j = 3$, $i = -1$.
$-1 = m_\gamma/(m_\alpha m_\beta) = n_\gamma/(n_\alpha n_\beta)$ on E
$0 = m_\delta/(m_\alpha^2 m_\beta) = n_\delta/(n_\alpha^2 n_\beta) + b - en_\gamma/(n_\alpha n_\beta) = n_\delta/(n_\alpha^2 n_\beta) + b + e$;
$n_\delta/(n_\alpha^2 n_\beta) = -b - e = -3$.
$0 = m_\epsilon/(m_\alpha^3 m_\beta) = n_\epsilon/(n_\alpha^3 n_\beta) - c + fn_\gamma/(n_\alpha n_\beta) - hn_\delta/(n_\alpha^2 n_\beta) =$
$n_\epsilon/(n_\alpha^3 n_\beta) - 1 + 3(-1) - 3(-3)$;
$n_\epsilon/(n_\alpha^3 n_\beta) = -5$.

$-1 = m_\zeta/(m_\alpha^3 m_\beta^2) = n_\zeta/(n_\alpha^3 n_\beta^2) - in_\epsilon/(n_\alpha^3 n_\beta)$
$-j(n_\gamma/(n_\alpha n_\beta))(n_\delta/(n_\alpha^2 n_\beta) = n_\zeta/(n_\alpha^3 n_\beta^2) + 1(-5) - 3(3)$;
$n_\zeta/(n_\alpha^3 n_\beta^2) = -1 + 5 + 9 = 13$.

Parts (c) and (d) now follow immediately. Write $(1-\alpha^{-1})/\lambda|_E = A, (1 - \beta^{-1})/\lambda|_E = B$, etc. By lemma 2.2,

$$\begin{aligned} w(\zeta) &= Z(n_\zeta/n_\alpha^3 n_\beta) + gA(n_\gamma/(n_\alpha n_\beta))^2 - iB(n_\epsilon/n_\alpha^3 n_\beta) - \\ &\quad j\Gamma(n_\gamma/n_\alpha n_\beta)(n_\delta/n_\alpha^2 n_\beta) - ejA(n_\gamma/n_\alpha n_\beta)^2 + ajA(n_\delta/n_\alpha^2 n_\beta) = \\ &= Z(13) + gA - iB(-5) - j\Gamma(-1)(-3) - ejA + ajA(-3) \\ &= 13Z + 3A - 5B - 9\Gamma - 6A - 9A \\ &= 13Z - 21A - 14B = 6Z \neq 0 \end{aligned}$$

because the tangent of our curve lies in no singular hyperplanes. This proves (e).

$w(\zeta)x(\eta) = w(\eta)x(\zeta)z(\alpha)^{m(\alpha)}z(\beta)^{m(\beta)}$ where $\zeta - \eta = m(\alpha)\alpha + m(\beta)\beta$. If $\eta \neq \zeta, x(\eta) = 0$ on E because $w(\zeta)$ is invertible, $m(\beta) > 0$, and $z(\beta)$ vanishes on E. Thus E is certainly not a subregular divisor.

IV. THE SUBREGULAR SPURIOUS DIVISOR

1. Subregular Unipotent Conjugacy Classes.

This section reviews well known results on the subregular conjugacy class of a reductive group. For details see [23]. Let G be a reductive group. A subregular unipotent conjugacy class of G is a conjugacy class whose centralizer has dimension $rank(G) + 2$. Conjugacy classes are taken over the algebraic closure of the base field unless specifically stated otherwise. Simple algebraic groups possess exactly one subregular unipotent conjugacy class. More generally a reductive group contains as many subregular unipotent conjugacy classes as there are connected components of the Dynkin diagram. Each subregular unipotent element determines a connected component of the Dynkin diagram. The variety of Borel subgroups containing a subregular unipotent element u, $(B\backslash G)_u$, is a union of projective lines and is called a Dynkin curve. Let P_α denote the parabolic subgroup associated with the simple root α. Each line is of the form $B\backslash P_\alpha g$ for some $g \in G$. The root α is uniquely determined by the line, and the line is said to be of type α. A line of type α does not intersect a line of type β if $(\alpha, \beta) = 0$. The number of lines in $(B\backslash G)_u$ and their incidence relations depends only on the connected component of the Dynkin diagram determined by u. There is always one line for each of the shorter roots, and two lines for each of the longer roots except for G_2 where there are three lines corresponding to the long root. There are $\langle -\beta, \alpha\rangle$ lines of type β intersecting each line of type α.

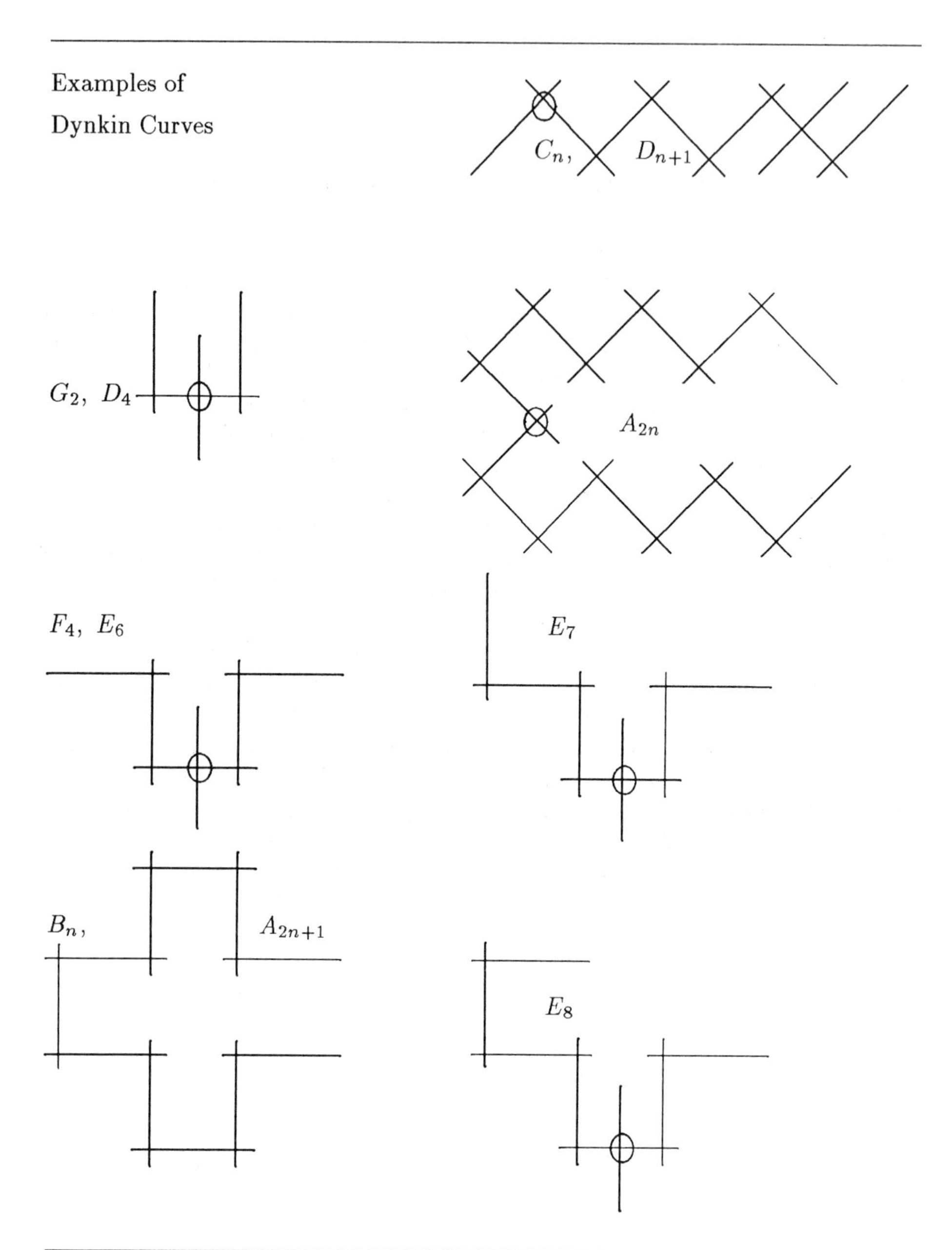

The line $B\backslash P_\alpha g$ is a line in $(B\backslash G)_u$ if and only if $u^{g^{-1}}$ is contained in the unipotent radical of P_α. For a point $p \in Y_\Gamma$ such that $\pi(p) = u$ is subregular unipotent, let $L_p(W)$ be the set of simple roots α such that $B(W)$ lies in a line of type α in $(B\backslash G)_u$. From the fact that no Borel subgroup lies in more than two lines of $(B\backslash G)_u$ it follows that $|L_p(W)| = 1$ or 2. Also if $\alpha, \beta \in L_p(W)$ then $(\alpha, \beta) \neq 0$ because a line of type α does not intersect any lines of type β when $(\alpha, \beta) = 0$. Consider a coordinate patch $S(B_\infty, B_0)$.

Recall $x(W,\alpha)$ is defined to be $x(\alpha)(u^{n_w^{-1}})$.

LEMMA 1.1. *$\alpha \in L_p(W)$ if and only if $x(W,\alpha) = 0$ at p.*

PROOF. $0 = x(W,\alpha)$ if and only if $u^{\nu^{-1}n_w^{-1}}$ lies in the unipotent radical of P_α the parabolic subgroup of type α containing the Borel B_0. So the line $B_0\backslash P_\alpha n_w \nu$ containing $B_0^{n_w\nu}$ lies in the Dynkin curve and conversely.

Note also that $L_p(W) = L_p(W')$ if $z(W,\alpha) = 0$ and that $z(W,\alpha) = 0$ if $B(W) = B(W')$ where (W,α) is the wall separating adjacent chambers W and W'. For every simple root α, we fix a wall (W_α,α) of type α such that $\theta_1(W_\alpha,\alpha) \neq 0$, and set

$$\tilde{z}(W,\alpha) = z(W,\alpha)/z(W_\alpha,\alpha) = z_1(W,\alpha)/z_1(W_\alpha,\alpha)$$

By the definitions of $\tilde{z}(W,\alpha)$ and $\theta_1(III.1)$ we have $\tilde{z}(W,\alpha) = 0$ if and only if $\theta_1(W,\alpha) = 0$. By abuse of language we will often say that the wall (W,α) is non-zero if $\theta_1(W,\alpha) \neq 0$. The following equations hold on $Y_1(B_\infty, B_0)$ for any two Weyl chambers W and W' where $T(W,\alpha) = (1-\gamma^{-1})/\lambda$ and $\pm\gamma$ is the root such that $\gamma = 0$ defines the wall (W,α).

EQUATION 1.2.

$$\lambda T(W,\alpha) = z(\alpha)z_1(W,\alpha)x(W,\alpha) = z(W,\alpha)x(W,\alpha)$$

$$\tilde{z}(W,\alpha)x(W,\alpha)T(W',\alpha) = T(W,\alpha)\tilde{z}(W',\alpha)x(W',\alpha).$$

$T(W,\alpha)$ is regular and invertible at $\lambda = 0$. Also it follows immediately from (1.1) and this equation that if $z(W,\alpha) \neq 0$ then $\alpha \in L_p(W)$.

COROLLARY 1.3. *For any given Weyl chamber W and p with $\pi(p) = u$, $u \in G(\bar{F})$ subregular there are at most two simple roots α such that $z(W,\alpha) \neq 0$. If $z(W,\alpha) \neq 0$ then $\alpha \in L_p(W)$.*

PROOF. No point of $(B\backslash G)_u$ lies in more than two lines.

2. Exclusion of Spurious Divisors.

THEOREM 2.1. *Let E be a subregular spurious divisor. Then $\beta(E) > 2$.*

PROOF. Select a wall (W,α) such that $\theta_1(W,\alpha) = 0$. Select a wall (W',α) such that $\theta_1(W',\alpha) \neq 0$. Form a path $W' = W_0, W_1, \dots, W_p = W$ from W' to W. Suppose that W_i and W_{i+1} are separated by a wall of type α_i, $i = 0, \dots, p-1$. Corresponding to this path is a sequence of walls (W_i,α) $i = 0, 1, \dots, p$. Let j be the smallest index for which $\theta_1(W_j,\alpha) = 0$. Since $\theta_1(W,\alpha) = 0$ such an index exists. Since $\theta_1(W',\alpha) \neq 0, j > 0$. Then $\theta_1(W_j,\alpha) = 0$ and $\theta_1(W_{j-1},\alpha) \neq 0$. The simple root α_{j-1} cannot equal α. By the nature of the equations holding at a node of type $A_1 \times A_1$ (III.1) we see that $(\alpha_{j-1},\alpha) \neq 0$.

Suppose first of all that α can be chosen so that $|\alpha_{j-1}| \geq |\alpha|$. The hypotheses of (III.3.1) now hold so that we can exclude the divisor.

Now assume that no matter how α is chosen $|\alpha_{j-1}| < |\alpha|$. It follows that if β is adjacent to α then $\theta_1(W,\beta) \neq 0$ for all W. Thus at the node defined by the two walls (W_{j-1},α_{j-1}) and (W_{j-1},α) none of the walls of type α_{j-1} are zero and $\theta_1(W_{j-1},\alpha) = 0, \theta_1(W_j,\alpha) \neq 0$. This by definition is a special node. Proposition III.1.1 now shows that $\beta(E) > 2$.

3. The graph Γ_0.

The remainder of this chapter discusses the structure of subregular spurious divisors and their zero patterns. These structural results will not be used in any of the following chapters. We assume that $G = A_n, B_n, C_n$ or $D_{n+1}, n \geq 3$ over an algebraically closed field $\bar{F}$. We fix a point p in a divisor such that $\pi(p) \in G(\bar{F})$ is subregular unipotent. Let R be the set of roots α such that $z(W,\alpha) \neq 0$ for some W. Let S be the set of roots α such that $\theta_1(W,\alpha) = 0$ for some W. Define a *solid node* to be a node at which $\theta_1(W,\alpha) \neq 0$ for all walls (W,α) at the node. We make the following assumption which remains in effect until section 8.

ASSUMPTION 3.1. $|S| \geq 2$ and if $\alpha, \beta \in S$ then there are no solid nodes of type (α,β).

We will see in section 8, that this assumption holds except in a few exceptional easily categorized cases.

LEMMA 3.2. $S \supseteq R$.

PROOF. Suppose $\alpha \in R \backslash S$. We will show that $|S| \leq 1$, contrary to assumption 3.1. If $\alpha \in R\backslash S$, then $z(W,\alpha) \neq 0$ for all W. Thus by (1.2), $x(W,\alpha) = 0$ for all W. Let $\beta \in S$. Then again by (1.2), $\tilde{z}(W,\beta)x(W,\beta) = 0$ for all β (since $\tilde{z}(W,\beta) = 0$ for some W). Pick $W = W'$ such that $\theta_1(W',\beta) \neq 0$. Then $x(W',\alpha)$ $= x(W',\beta) = 0$. Since $\pi(p)$ is assumed to be subregular, α and β are adjacent.

Assume that α and β are the same length. Then the fact that $\theta_1(W,\alpha) \neq 0$ for all (W,α) together with the fact that $\theta_1(W',\beta) \neq 0$ forces the node of type A_2 formed by the walls (W',α) and (W',β) to be a solid node. (See the list of zero patterns for a node of type A_2.) Thus

$$x(W,\alpha) = x(W,\beta) = 0$$

for all walls (W,α), (W,β) at the node. This shows that $\alpha, \beta \in L_p(W)$ for all Weyl chambers W at the node. There is exactly one Borel subgroup B_+ which lies at the intersection of a line of type α with a line of type β. Thus $B(W) = B_+$ for all W at the node. This contradicts the fact that $z(W',\alpha) \neq 0$ $(B(W') \neq B(W''))$ where W'' is the chamber through a wall of type α from W'). We conclude that α and β have different lengths.

For a reductive group the assumption that S is a set such that if $\beta \in S$ then $(\beta,\alpha) \neq 0$ and α and β are different lengths forces $|S| \leq 1$. This proves the lemma.

We form an equivalence relation on the non-zero walls (W,α) of type α for $\alpha \in S$. Say that two non-zero walls (W,α) and (W',α) are equivalent if (I) there is a path $W = W_0, \ldots, W_p = W'$ from W to W' such that if $\theta_1(W_i,\alpha_i) \neq 0$ then

$\alpha_i \notin S$ or if (II) (W, α) and (W', α) are the two non-zero colinear walls at a node of type B_2 with zero pattern $B.III$ or $B.IV$:

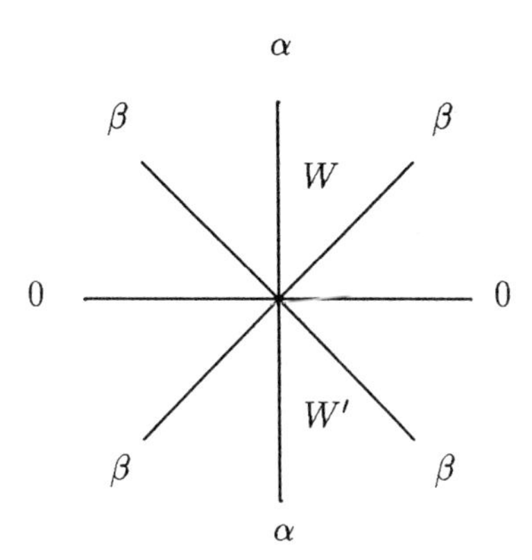

The equivalence classes are defined to be the smallest classes such that (I) and (II) are satisfied.

Form a graph Γ_0 as follows. Let each vertex be given by an equivalence class of walls. A vertex v_1 of Γ_0 is said to be of type α ($type(v_1) = \alpha$) if the walls of the equivalence class are of type α. Join two vertices ($v_1 \neq v_2$) by an edge if there is a series of chambers $W_0, \ldots, W_p$ separated by walls (W_i, α_i) $i = 0, \ldots, p-1$ such that $(W_0, type(v_1))$ is a wall of $v_1, (W_p, type(v_2))$ is a wall of v_2, and if $\theta_1(W_i, \alpha_i) \neq 0$ then $\alpha_i \notin S$ for $i = 0, \ldots, p-1$.

LEMMA 3.3. *If the vertices v_1 and v_2 are joined then*

$$(type(v_1), type(v_2)) \neq 0.$$

PROOF. By construction a vertex is not joined to itself by an edge.

We use equation 1.2 in the form

$$\tilde{z}(W, \alpha)x(W, \alpha)T(W', \alpha) = T(W, \alpha)\tilde{z}(W', \alpha)x(W', \alpha).$$

Since $type(v_1) \in S$, there is a W' such that $\theta_1(W', type(v_1)) = 0$. It follows that $\tilde{z}(W, \alpha)x(W, \alpha) = 0$ for all W where $\alpha = type(v_1)$. If $(W, \alpha) \in v_1$ then $\theta_1(W, \alpha) \neq 0$ and $x(W, \alpha) = 0$ so that $\alpha \in L_p(W)$. Similarly, if

$$(W', type(v_2)) \in v_2 \quad \text{then} \quad type(v_2) \in L_p(W').$$

Selecting a path between appropriately selected chambers W and W', we see that the intervening walls are zero ($z(W_i, \alpha_i) = 0$)(3.2) so that $B(W) = B(W')$ and $L_p(W) = L_p(W')$. So

$$type(v_1),\ type(v_2) \in L_p(W).$$

$B(W)$ lies in lines of type $type(v_1)$ and $type(v_2)$. By the nature of the Dynkin curve $(type(v_1), type(v_2)) \neq 0$.

Every path $W_0, \ldots, W_p$ through the Weyl chambers gives rise to a path through the graph Γ_0 as follows. Let $W_{a_1}, \ldots, W_{a_j}$ $a_1 < a_2 < \ldots < a_j$ be the indices of the chambers in the path such that $\theta_1(W_{a_i}, \alpha_{a_i}) \neq 0$ and $\alpha_{a_i} \in S$. Then let v_i be the vertex corresponding to the wall (W_{a_i}, α_{a_i}). By construction v_i is joined by an edge to v_{i+1} provided $v_i \neq v_{i+1}$. It is clear that if the path through the Weyl chambers is taken to be closed $W_0, \ldots, W_p, W_{p+1} = W_0$ then the corresponding path through the graph Γ_0 will be closed. Since we can always select a path $W_0, \ldots, W_p$ between any two given walls (W, α) and (W', β), there is a path in Γ_0 between any two given vertices. Thus Γ_0 must be a connected graph. This proves part (a) of the following lemma.

LEMMA 3.4.
a) The graph Γ_0 *is connected.*
b) Γ_0 *is a tree.*
c) S *forms a connected Dynkin diagram.*

PROOF. (a) has been proved and (c) follows directly from (a) and (3.3).

(b) We have described a map from closed paths $W_0, \ldots, W_p, W_{p+1} = W_0$ through the Weyl chambers to the closed paths in Γ_0. (b) will follow from two facts. First, a homotopy of paths in the Weyl chambers maps to a homotopy of paths in Γ_0. Second, every closed path in Γ_0 is homotopic to the image of a closed path through the Weyl chambers. If we have these two facts then (b) follows from the fact that any closed path through the Weyl chambers is contractable. We do not require that the homotopies fix a base point.

To check that a homotopy of paths in the Weyl chambers maps to a homotopy of paths in Γ_0, it is enough to check the statement as the homotopy passes through a wall or node. We begin with a special case. Suppose that part of the path L is

$$\ldots W_i, W_{i+1}, \ldots, W_{i+k} = W_i, W_{i+k+1}, \ldots$$

which a homotopy carries to

$$L' : \ldots W_i, W_{i+k+1}, \ldots$$

Suppose that there is a vertex v of Γ_0 such that for all j in the range $i \leq j < i+k$ we have: If $\theta_1(W_j, \alpha_j) \neq 0$ and $\alpha_j \in S$ then (W_j, α_j) belongs to the vertex v of Γ_0. Let (W_ℓ, α_ℓ) be the wall with the largest index ℓ such that $\ell < i, \alpha_\ell \in S$ and $\theta_1(W_\ell, \alpha_\ell) \neq 0$. Let v_a be the vertex of Γ_0 corresponding to (W_ℓ, α_ℓ). If $v \neq v_\alpha, v$ and v_α are joined by an edge. Similarly let $(W_{\ell'}, \alpha_{\ell'})$ be the wall with the smallest index ℓ' such that $\ell' \geq i+k, \alpha_{\ell'} \in S$ and $\theta_1(W_{\ell'}, \alpha_{\ell'}) \neq 0$. Let v_b be the vertex of Γ_0 corresponding to $(W_{\ell'}, \alpha_{\ell'})$. If $v \neq v_b$ then v and v_b are joined by an edge.

I claim that v_a, v_b, and v are not distinct vertices. For if they were distinct, there would be edges between each pair, and their types (a simple root in S) would be distinct. This by (3.3), would give a closed loop in the Dynkin diagram of G.

If $v = v_a$ or v_b then paths L and L' give rise to the same path in Γ_0. If $v_a = v_b$, then the paths in Γ_0 corresponding to L and L' are given by the following diagram.

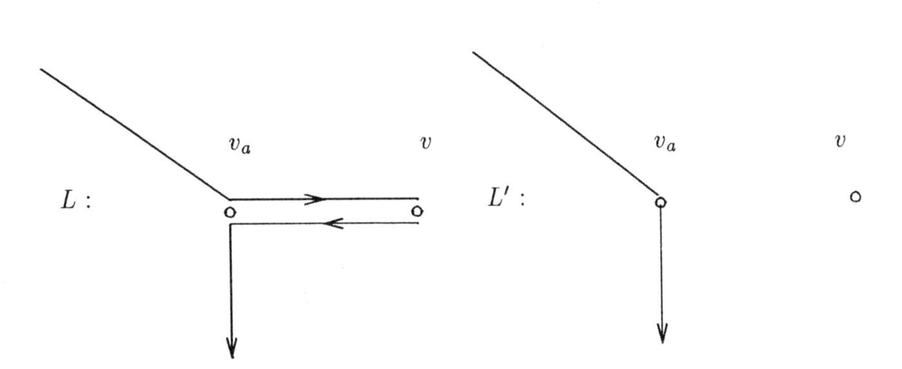

They are clearly homotopic.

This special case takes care of the situation where the homotopy moves through a wall or through a node of type (α, β) such that if $\beta \in S$ then $\theta_1(W, \beta) = 0$ for all walls (W, β) at the node. We must still consider a homotopy that moves through a node of type (α, β) with $\alpha, \beta \in S$. By assumption 3.1, we may assume that it is a special node. The following diagrams now make it clear that the homotopy of paths in the Weyl chambers translates to a homotopy of Γ_0.

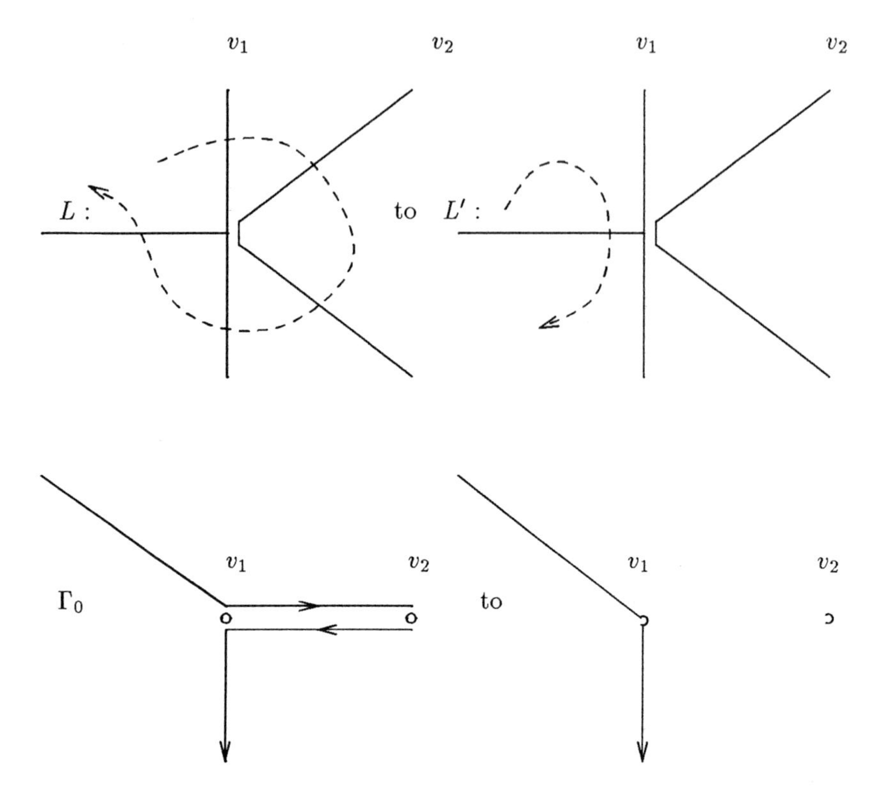
v_1
v_2
v_1
v_2
L :
to
L' :
v_1
v_2
Γ_0
to
v_1
v_2

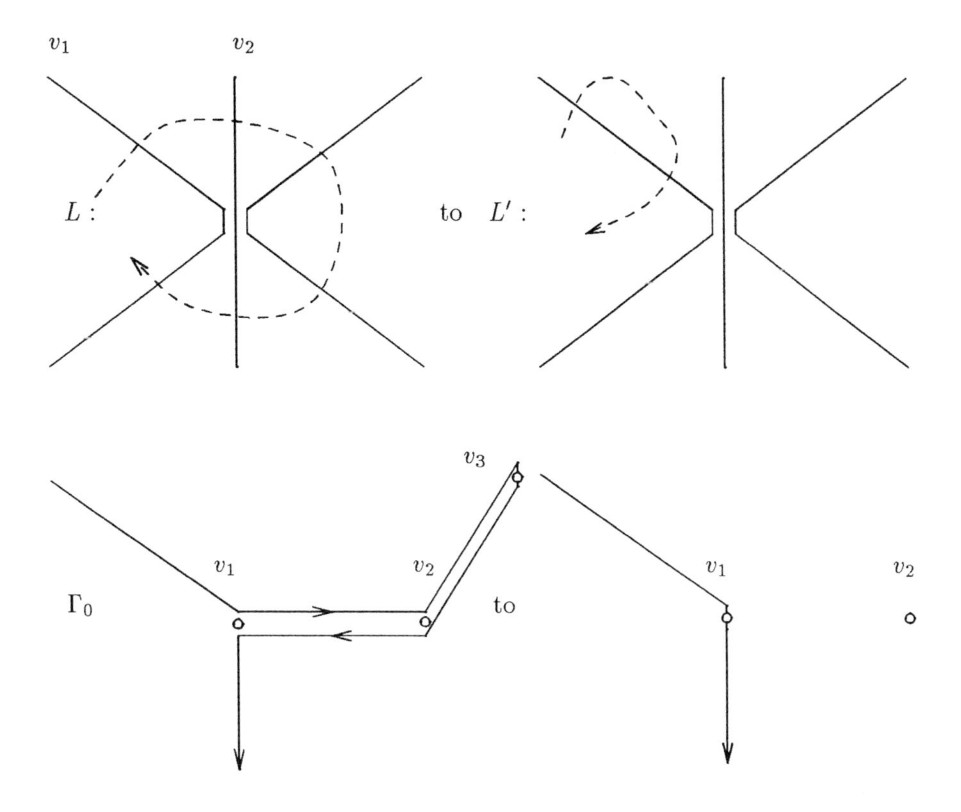

We turn to the second fact: that every path in Γ_0 has the same homotopy class as the image of a path through the Weyl chambers. This would be obvious by the definitions of the vertices and edges of the graph were it not for condition (II) defining the equivalence relation on walls belonging to a given vertex. It is enough to check that if (W, α) and (W', α) are the two non-zero colinear walls at a node (with pattern B.III or B.IV) belonging to the vertex v of Γ_0, then there is a path joining (W, α) to (W', α) whose image in Γ_0 is a path homotopic to the constant path at the vertex v. This is completely evident from the following diagram.

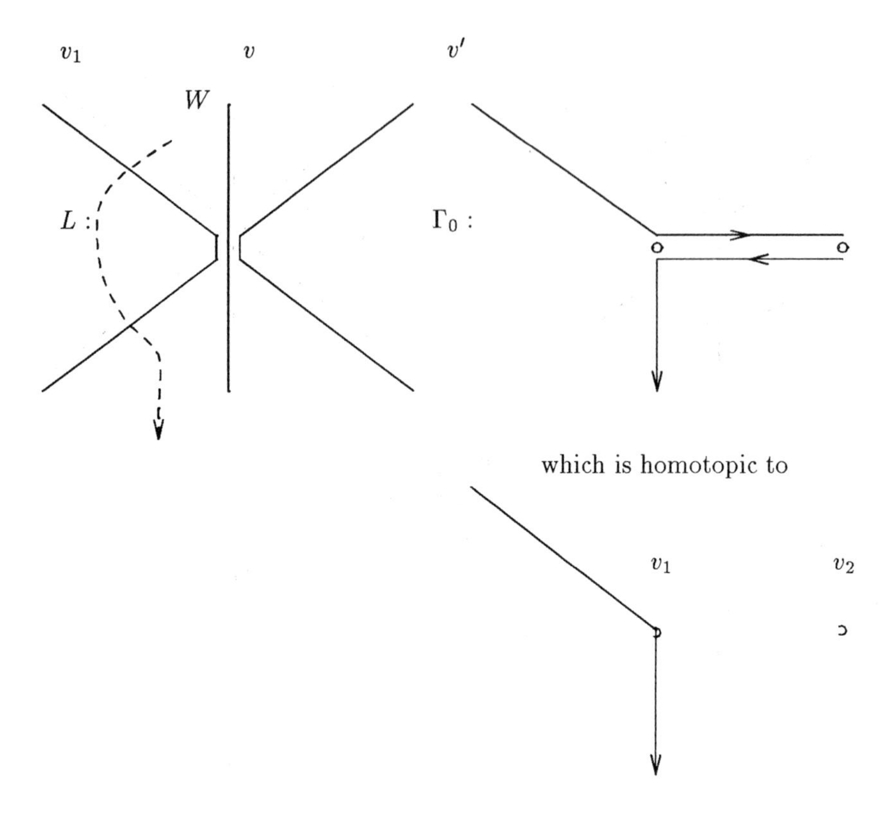

4. The Modified Star.

Let S_2 be the affine variety generated by the variables $\hat{z}(\alpha)$ $\forall\, \alpha$, $\hat{z}(W,\alpha)$ $\forall\, (W,\alpha)$ subject to the relations:

i) $\hat{z}(W,\alpha) + \hat{z}(W'\alpha) = 0$ where W and W' are adjacent walls separated by a wall of type α, and

ii) $\exp(\hat{z}_p X_{-\alpha_p}) \ldots \exp(\hat{z}_1 X_{-\alpha_1}) = 1$ where $W_1, W_2, \ldots, W_p$ is the path around a node (so that $p = 4, 6, 8$ according as the node is of type $A_1 x A_1, A_2, B_2$) and $\hat{z}_1, \ldots, \hat{z}_p$ are the corresponding variables.

iii) $\hat{z}(W_\alpha, \alpha) = 1$ for some chamber W_α (unless $\hat{z}(W,\alpha) = 0, \quad \forall W$).

By (I.3.1) there is an injection from the affine patch

$$\{x \in S_1(B_\infty, B_0) : \theta_1(W_\alpha, \alpha) \neq 0\}$$

to S_2 given by $\hat{z}(W,\alpha) \to \tilde{z}(W,\alpha)$, $\hat{z}(\alpha) \to z(W_\alpha, \alpha)$. Elements of S_2 will be called *modified stars.*

LEMMA 4.1. *Suppose that all the walls corresponding to a given vertex of* Γ_0 *are set equal to zero. Then the resulting equations define a modified star.*

PROOF. To check this it is sufficient to verify that the equations at every node are still satisfied. But this is immediate from the definition of a vertex of Γ_0.

4.2 Modifications.

We now modify the graph. Every time a vertex is eliminated from the graph, we set all corresponding walls to zero in the star. Doing so gives a new modified star by (4.1). The new star will have a graph Γ' associated to it. When an extremal vertex of the graph Γ_0 is eliminated, the modified graph Γ' is a subgraph of Γ_0. If $G = D_n, |S| \geq 4, \alpha_+, \alpha_- \in S$ where α_+ and α_- are simple roots interchanged by an outer automorphism of G, then let $S_- = S \backslash \{\alpha_-\}$. Otherwise set $S_- = S$. Thus S_- forms a connected Dynkin diagram with two extremal simple roots.

Modify the graph as follows.

1. For every extremal simple root of the Dynkin diagram S_- fix a vertex of that type in Γ_0.
2. Let Γ_1 be the minimal tree containing these two vertices. Eliminate the vertices not in Γ_1 and set the corresponding walls of the star equal to zero. The resulting star will have graph Γ_1.
3. Γ_1 is linear. Let $\alpha_1, \dots, \alpha_n$ be the roots of S_- ordered in such a way that if $(\alpha_i, \alpha_j) \neq 0$ then $|i - j| \leq 1$. Let v_1 (resp. v_n) be the extremal vertex in Γ_1 corresponding to α_1 (resp. α_n). Order the vertices in Γ_1 from $u_1 = v_1$ up to $u_k = v_n$. For each i $(1 \leq i \leq n)$ eliminate all vertices of type α_i except for the last. The corresponding star ($\in S_2$) has one equivalence class of type α_i for all i. Its graph Γ' is the Dynkin diagram S_-.

The modified star will be easier to work with because its graph is the same as the Dynkin diagram S_-.

The non-zero walls of a given type α of the modified star divide the Weyl chambers into *α-regions* each region being a union of Weyl chambers bounded by non-zero walls of type α. More precisely, define an equivalence relation on the Weyl chambers by making two chambers equivalent if there is a path from one to the other which does not pass through any non-zero walls of type α. Each region is then defined to be the union of the chambers in an equivalence class.

LEMMA 4.2. *Suppose $\alpha, \beta \in S$ and $\alpha \neq \beta$. Then there is a β-region inside which all the non-zero walls of type α lie.*

PROOF. This lemma will follow immediately from definitions if we show that in the modified star there are no nodes with zero pattern $B.III$ or $B.IV$. In the graph Γ' of the modified star there is only one vertex of each type. Thus we must show that the walls $(W, \beta), (W', \beta)$ of type β on opposite sides of the non-zero walls $(W, \alpha) = (W', \alpha)$ of type α are inequivalent. Suppose there were a path $W = W_0, \dots, W_p = W'$ from W to W' such that $\theta_1(W_i, \alpha_i) = 0$ if $\alpha_i = \alpha$. Then the condition that the non-zero walls of a given type in a closed loop sum to zero gives $\hat{z}(W, \alpha) = 0$. This is a contradiction.

REMARK 4.3. The argument of the lemma also shows that the α-regions have no *"internal"* non-zero walls of type α.

5. The Weyl Chambers.

To obtain more detailed information about the graph Γ_0 we must first study the Weyl chambers. Weyl chambers are taken to be closed rather than open sets in $\mathbb{R}^n$, so that the walls of a Weyl chamber are contained in the chamber.

For a fixed simple root α define an equivalence relation on the Weyl chambers as follows. Two chambers W, W' are equivalent if and only if W and W' are connected by a path $W = W_1, \ldots, W_p = W'$ such that for $i = 1, \ldots, p-1$ the chambers W_i and W_{i+1} are adjacent chambers, and the wall separating them is not type α. The union of all chambers in an equivalence class is called an *α-chamber* or a *big chamber* of type α. Each α-chamber is bounded by walls of type α. Similarly, given a set Q of simple roots, big chambers of type Q can be defined. If a wall of type α bounds a big chamber of type Q then $\alpha \in Q$.

The union of all walls of type α which lie entirely in the intersection of an α-chamber with a hyperplane in $\mathbb{R}^n$ is called a *big wall* or *α-wall* . Let ξ_1 and ξ_2 be connected components of two α-walls such that the intersection of ξ_1 and ξ_2 is codimension two. Then the intersection is a union of nodes. Such walls are said to be *adjacent* . The angle formed by the walls is $\pi/2$ (nodes of type B_2) or $2\pi/3$ (nodes of type A_2). In the latter case but not the former the big walls are said to be *obtusely adjacent.* Since the angle between adjacent walls is always less than or equal to π, each big chamber is convex. Therefore each big chamber $\mathbb{W}$ is defined by the intersection of a finite number of half spaces $E_1, \ldots, E_q$. The big walls are $\partial E_i \cap \mathbb{W}$ and are therefore convex and thus connected. The codimension two intersections are called *big nodes.* Either three big walls come together at angles of $2\pi/3$ or four big walls come together at angles of $\pi/2$. From this it is clear that each big wall must bound exactly two big chambers. Since a big wall lies in a hyperplane, the only nodes between walls forming a big wall must be of type $A_1 \times A_1$. By the connectivity of the big walls and the nature of the equations holding at a node of type $A_1 \times A_1 (III.1)$ it follows that $\hat{z}(W, \alpha) = \hat{z}(W', \alpha)$ for all walls $(W, \alpha), (W', \alpha)$ forming a big wall.
Label the roots as follows:

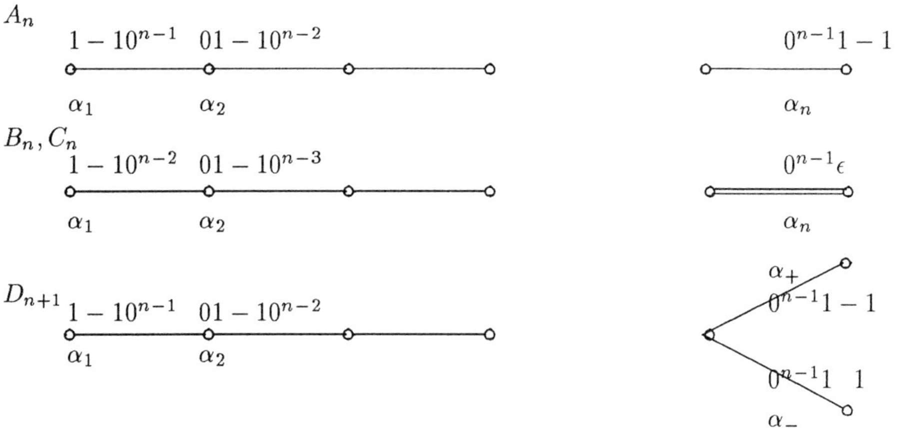

A_n

$1-10^{n-1}$ $\quad$ $01-10^{n-2}$ $\qquad$ $0^{n-1}1-1$

α_1 $\quad$ α_2 $\qquad$ α_n

B_n, C_n

$1-10^{n-2}$ $\quad$ $01-10^{n-3}$ $\qquad$ $0^{n-1}\epsilon$

α_1 $\quad$ α_2 $\qquad$ α_n

D_{n+1}

$1-10^{n-1}$ $\quad$ $01-10^{n-2}$ $\qquad$ α_+ $\quad$ $0^{n-1}1-1$

α_1 $\quad$ α_2 $\qquad$ $0^{n-1}1\ 1$ $\quad$ α_-

Define the *fundamental α_1-cell* to be the α_1-chamber containing the fundamental Weyl chamber. Define the *fundamental α_{i+1}-cell* $i = 1, \ldots, n-1$ to be the smallest union of α_{i+1}-chambers to contain the fundamental α_i-cell (*i.e.*

the union of all α_{i+1}-chambers meeting the interior of the fundamental α_i-cell). Define an α_i*-cell* to be a translate of the fundamental α_i-cell by the Weyl group.

LEMMA 5.1. *Given any α_p-chamber $\mathbb{W}$ and a wall ξ of $\mathbb{W}$ there exists an α_p-cell containing ξ but not $\mathbb{W}$.*

PROOF. Let W be a Weyl chamber with wall (W, α_p) contained in α_p-chamber $\mathbb{W}$ and α_p-wall ξ respectively. Let W' be a Weyl chamber which is not contained in the fundamental α_p-cell but whose wall (W', α_p) is. A Weyl group element takes W' to $W, (W', \alpha_p)$ to (W, α_p) and the fundamental cell to an α_p-cell K. Then K docs not contain $\mathbb{W}$ but does contain ξ.

Now fix a simple root $\alpha = \alpha_p$ where $1 \leq p \leq n$ ($1 \leq p \leq n-1$ for D_{n+1}). Let $L = \overline{K^c}$ be the closure of the complement of the fundamental α-cell. Call a big wall of L *external* if it lies in the intersection of L and K. L can be broken up into two sets $L^+ \cup L^-$. Each is a union of α-chambers. The set L^+ is the union of α-chambers in L which contain an external wall. L^- is the union of all other α-chambers in L. The rest of the section establishes the following facts about the structure of L.

FACT 1. If $\mathbb{W}$ is an α-chamber in L^-, then $\mathbb{W} \cap L^+$ contains a big wall.

FACT 2. (Assume $\alpha \neq \alpha_+, \alpha_-$ for D_{n+1}). If $p > 1$, any two obtusely adjacent external walls which lie in the same α-chamber $\mathbb{W} \subseteq L^+$ meet at a big node which contains a node of type (α_p, α_{p-1}). (Recall that a big node is a union of nodes).

FACT 3. (Assume $\alpha \neq \alpha_n$ for $B_n, C_n; \alpha \neq \alpha_+, \alpha_-$ for D_{n+1}). The external walls of L lying in a given α-chamber $\mathbb{W}$ are connected in the following sense. Given two external walls ξ, ξ' in $\mathbb{W}$ there exists a chain of external walls $\xi = \xi_1, \ldots, \xi_t = \xi'$ in $\mathbb{W}$ such that ξ_i, ξ_{i+1} are obtusely adjacent for $i = 1, \ldots, t-1$.

FACT 4. (Assume $\alpha \neq \alpha_n$ for B_n,C_n; $\alpha \neq \alpha_+, \alpha_-$ for D_{n+1}). Let $\mathbb{W}$ and $\mathbb{W}'$ be adjacent α-chambers in L^+ separated by a big wall ξ''. Then there are external walls ξ in $\mathbb{W}$ and ξ' in $\mathbb{W}'$ such that ξ, ξ', ξ'' form a big node with angles $2\pi/3$.

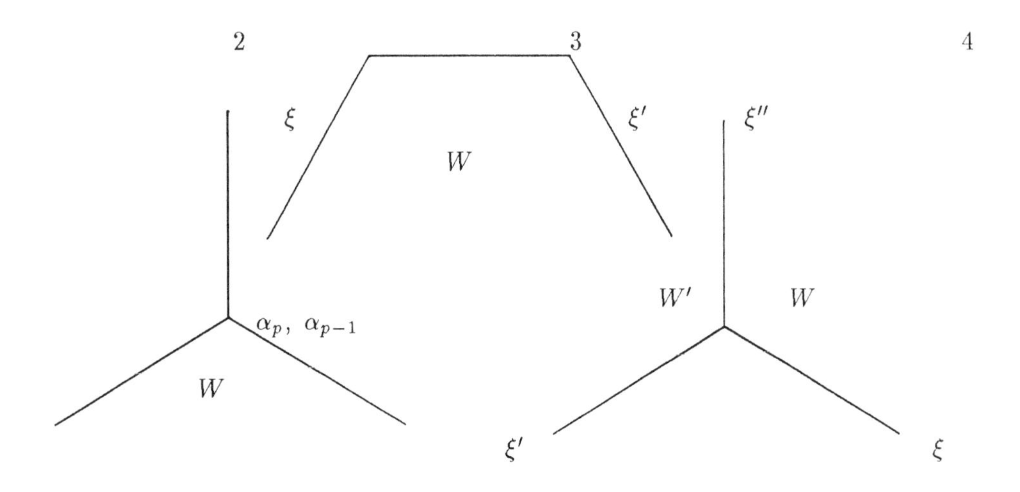

The Weyl Chambers.

In the following we must discuss particular cases.

$\underline{A_n}$. Let the Weyl chambers lie in the hyperplane $\mathbf{P} = \{x \in \mathbb{R}^{n+1} : \sum x_i = 0\}$. The fundamental chamber W_+ is given by

$$W_+ = \{x \in \mathbf{P} : (x, \alpha) \geq 0 \quad \text{for all } \alpha\} = \{x \in \mathbf{P} : x_1 \geq x_2 \geq \ldots \geq x_{n+1}\}.$$

The Weyl group acts on $\mathbf{P}$ by permuting the coordinate axes. So $|\Omega| = (n+1)!$ Each chamber has n walls. If $\mathbf{e}_1, \ldots, \mathbf{e}_{n+1}$ is a standard basis of $\mathbb{R}^{n+1}$ then $\pi \in \Omega$ acts on $\{1, \ldots, n+1\}$ by $\pi(\mathbf{e}_i) = \mathbf{e}_{\pi i}$. Then

$$\begin{aligned} W(\pi) =& \pi^{-1} W_+ \\ =& \{x \in \mathbf{P} : x_{\pi_1^{-1}} \geq x_{\pi_2^{-1}} \geq \ldots \geq x_{\pi_j^{-1}} \geq x_{\pi_{j+1}^{-1}} \geq \ldots \geq x_{\pi_{n+1}^{-1}}\}; \end{aligned}$$

and if w_j equals the permutation $(j, j+1)$ then

$$W(w_j \pi) = \{x \in \mathbf{P} : x_{\pi_1^{-1}} \geq x_{\pi_2^{-1}} \geq \ldots \geq x_{\pi_{j+1}^{-1}} \geq x_{\pi_j^{-1}} \geq \ldots \geq x_{\pi_{n+1}^{-1}}\}.$$

So the wall of type α_j of $W(\pi)$ is given by

$$\{x \in W(\pi) : x_{\pi_j^{-1}} = x_{\pi_{j+1}^{-1}}\}.$$

Fix a simple root $\alpha = \alpha_p$. Then the α_p-chamber containing W_+ is obtained by reflecting through walls other than α_p. In terms of the defining equations the α_p-chamber is obtained by interchanging the kth variable with the $k+1^{st}$ variable for $k \neq p$. This shows that the α_p-chamber is

$$\{x \in \mathbf{P} : \min_{i \leq p}(x_i) \geq \max_{i > p}(x_i)\}.$$

To facilitate the description of the cells define

$$rank(x, i) = 1 + |\{j : x_j > x_i\}|$$

for $x \in \mathbb{R}^{n+1}$ and $i \in \{1, \ldots, n+1\}$. It is the order in which the variable x_i occurs among the variables $x_1, \ldots, x_{n+1}$ when they are ranked according to size. The α_p-chamber containing W_+ becomes

$$\{x \in \mathbf{P} : rank(x, i) \leq p, i = 1, \ldots, p\}.$$

The fundamental α_1-cell is

$$\{x \in \mathbf{P} : rank(x, 1) \leq 1\}.$$

The α_2-chambers which meet this are

$$\{x \in \mathbf{P} : rank(x, i) \leq 2, \quad i = 1, j\}$$

for $j = 2, \ldots, n+1$. So the fundamental α_2-cell is

$$\{x \in \mathbf{P} : rank(x,1) \leq 2\}.$$

Inductively, we obtain that the fundamental α_p-cell equals

$$\{x \in \mathbf{P} : rank(x,1) \leq p\}.$$

Acting on the fundamental α_p-cell by the Weyl group we obtain the α_p-cells

$$W(p,i) = \{x \in \mathbf{P} : rank(x,i) \leq p\}.$$

Notice that an α_n-cell is the closure of the complement of an α_1-cell. The set L introduced above is given by the closure of $\{x \in \mathbf{P} : rank(x,1) > p\}$. The α_p-chambers in L are

$$\mathbb{W} = \{x \in \mathbf{P} : rank(x,i) \leq p, i \in I\}$$

for all sets I $\subseteq \{2, \ldots, n+1\}$ of cardinality p. Let W be a Weyl chamber in $\mathbb{W}$ with $rank(x,i) \leq p$, $i \in I$ and $rank(x,1) = p+1$ in the interior of W. Then (W, α_p) forms a part of an external wall. So L^- is the empty set and fact 1 is trivial.

Consider the α_p-chamber $(p > 1)$

$$\mathbb{W} = \{x \in \mathbf{P} : rank(x,i) \leq p, i = 2, \ldots, p+1\}.$$

It lies in L. The external walls are given by $\{x \in \mathbb{W} : x_i = x_1\}$ $i = 2, \ldots, p+1$. Consider the walls $x_j = x_1$ and $x_i = x_1$ $(i,j$ fixed $2 \leq i,j \leq p+1)$. Let W be a Weyl chamber in $\mathbb{W}$ with $rank(x,j) = p-1$, $rank(x,i) = p$, $rank(x,1) = p+1$ in the interior of W. Then the walls (W, α_{p-1}) $(x_i = x_j)$ and (W, α_p) $(x_i = x_1)$ are walls of an (α_{p-1}, α_p) node forming a part of a big node at which the walls $x_j = x_1$ and $x_i = x_1$ meet. This gives facts 2 and 3.

If $p = n, L$ consists of a single α_p-chamber and fact 4 is trivial. So suppose that $p < n$. Consider adjacent α_p-chambers $\mathbb{W}$, $\mathbb{W}'$ in L separated by ξ. By relabeling if necessary we have

$$\begin{aligned}
\mathbb{W} &= \{x \in \mathbf{P} : rank(x,i) \leq p, i = 2, \ldots, p+1\}, \\
\mathbb{W}' &= \{x \in \mathbf{P} : rank(x,i) \leq p, i = 3, \ldots, p+2\}, \\
\xi &= \{x \in \mathbb{W} \cap \mathbb{W}'\} = \{x \in \mathbb{W} : x_3 = x_{p+2}\}.
\end{aligned}$$

An external wall is given by $\xi' = \{x \in \mathbb{W} : x_3 = x_1\}$. They meet at a node of type (α_p, α_{p+1}) along with the wall

$$\xi'' = \{x \in \mathbb{W}' : x_1 = x_{p+2}\}.$$

This gives fact 4.

$\mathbf{B_n}, \mathbf{C_n}$. The Weyl chambers lie in $\mathbf{P} = \mathbb{R}^n$. The fundamental chamber W_+ is given by

$$\{x \in \mathbf{P} : (x, \alpha) \geq 0\} = \{x \in \mathbf{P} : x_1 \geq x_2 \geq \ldots \geq x_n \geq 0\}.$$

The Weyl group acts on $\mathbf{P}$ by permuting the coordinate axes and changing signs. So $|\Omega| = 2^n n!$ Each chamber has n walls. Define a modified rank function by $rank'(x,i) = rank(x',i)$ where $x = (x_1, \ldots, x_n)$ and $x' = (|x_1|, \ldots, |x_n|)$. The α_p-chamber containing W_+ ($p \neq n$) is obtained by interchanging x_i and x_{i+1} ($i \neq p$) and negating x_j ($j > p$). This gives

$$\{x \in \mathbf{P} : \min_{i \leq p}(x_i) \geq \max_{i > p} |x_i|\}$$

as the α_p-chamber containing W_+. In terms of the rank function it is

$$\{x \in \mathbf{P} : rank'(x,i) \leq p, \quad i = 1, \ldots, p \quad \text{and} \quad x_i \geq 0, i = 1, \ldots, p\}.$$

The fundamental α_p-cell is

$$\{x \in \mathbf{P} : rank'(x,1) \leq p, x_1 \geq 0\}.$$

The other α_p-cells are

$$W_\pm(p,i) = \{x \in \mathbf{P} : rank'(x,i) \leq p, \pm x_i \geq 0\}.$$

Note that $W_\pm(n,i)$ is a half space.

A simplifying notation can be introduced for big chambers. If $\mathbb{W}$ is an α_p-chamber it is specified by a subset $J \subseteq \{1, \ldots, n\}$ with $|J| = p$ and a function $\epsilon : J \to \{\pm 1\}$. Let $\mathbb{W}\{\epsilon, J\}$ be the α_p-chamber

$$\{x \in \mathbf{P} : rank(x,i) \leq p, \epsilon_i x_i \geq 0 \quad \text{for all} \quad i \in J\}.$$

If $\mathbb{W}$ is a big chamber ($\mathbb{W} = \mathbb{W}\{\epsilon, J\}$), let $\mathbb{W}[a_1, \ldots, a_k | b_1, \ldots, b_\ell]$ (where $a_i, b_i \in \{\pm 1, \ldots, \pm n\}; i = 1, \ldots, k; j = 1, \ldots, \ell$ and $|a_1|, \ldots, |a_k|, |b_1|, \ldots, |b_\ell|$ are all distinct) be the big chamber given by $J' = (J \cup \{|a_i|\}) \backslash \{|b_i|\}$ and $\epsilon'(x) = sign(a_i)$ if $x = |a_i|$ for some a_i, and $\epsilon'(x) = \epsilon(x)$ otherwise. If $\mathbb{W}$ is an α_p-chamber $p \neq n$ we let $\mathbb{W}\langle a, b\rangle$ ($a, b \in \{\pm 1, \ldots, \pm n\}$) be the wall of $\mathbb{W}$ given by

$$sign(a) x_{|a|} = sign(b) x_{|b|}$$

where exactly one of $|a|, |b|$ lies in J and $sign(a) = \epsilon(a)$ if say $a \in J$. Also when $p \neq n$ let $\mathbb{W}\langle a, b, c\rangle$ be the node of $\mathbb{W}$ given by

$$sign(a) x_{|a|} = sign(b) x_{|b|} = sign(c) x_{|c|}$$

where either one or two of $|a|, |b|, |c|$ lie in J, and $sign(r) = \epsilon(r)$ for

$$r \in \{|a|, |b|, |c|\} \cap J.$$

With this notation it is clear that the big chambers at the node $\mathbb{W}\langle a,b,c\rangle$ are $\mathbb{W}[a,b|c]$, $\mathbb{W}[b,c|a]$, $\mathbb{W}[c,a|b]$, $\mathbb{W}[a|b,c]$, $\mathbb{W}[b|c,a]$, and $\mathbb{W}[c|a,b]$. When we write an equation such as $\mathbb{W} = \mathbb{W}[|a]$ we mean that $\mathbb{W}$ is a big chamber such that $|a|\notin J$.

If $\mathbb{W}$ is an α_p-chamber that does not lie in the fundamental α_p-cell then either $\mathbb{W} = \mathbb{W}[-1|]$ or $\mathbb{W} = \mathbb{W}[|1]$. In the latter case $\mathbb{W}\langle j|1\rangle$ is an external wall so these $\mathbb{W}$ must lie in L^+. In the former case $\mathbb{W}\langle j,-1\rangle$ is a big wall in $\mathbb{W}\cap L^+$ for all j (provided $p \neq n$). If $p = n$, $\mathbb{W} = \mathbb{W}[-1|]$ shares the wall $x_1 = 0$ with $\mathbb{W}[1|]$ so that $L^- = \phi$. This proves fact 1.

Fact 2 is trivial for B_n, C_n if $p = n$. So for facts 2,3 $p \neq 1, n$. Consider an α_p-chamber $\mathbb{W} = \mathbb{W}[|1]$ and the external walls $\mathbb{W}\langle j,1\rangle$, $\mathbb{W}\langle j',1\rangle$ $(|j| \neq |j'|)$. They meet at the node $\mathbb{W}\langle 1,j,j'\rangle$ of type (α_p, α_{p-1}). If $p = 1$ $\mathbb{W} = \mathbb{W}[i|1]$ has only one external wall $\mathbb{W}\langle i,1\rangle$ and fact 3 is trivial.

When $p = n-1$, $\mathbb{W}$ is never adjacent to $\mathbb{W}'$ if $\mathbb{W}$, $\mathbb{W}'\subseteq L^+$ for $\mathbb{W} = \mathbb{W}[|1]$ can only pass to $\mathbb{W}[1|x]$ or $\mathbb{W}[-1|x]$. So assume $p \leq n-2$. Let $\mathbb{W} = \mathbb{W}[x|1,y]$, $\mathbb{W}' = \mathbb{W}[y|1,x]$. Then we have the node $\mathbb{W}\langle 1,x,y\rangle$ with external walls $\mathbb{W}\langle 1,x\rangle$, $\mathbb{W}'\langle 1,y\rangle$. This proves fact 4.

$\mathbf{D_{n+1}}$ The fundamental Weyl chamber is

$$W_+ = \{x \in \mathbf{P} : (x,\alpha) \geq 0\} = \{x \in \mathbf{P} : x_1 \geq \ldots \geq x_n \geq |x_{n+1}|\}.$$

The Weyl group permutes coordinate axes and changes signs of all but the smallest variable so $|\Omega| = 2^n(n+1)!$ Each chamber has $n+1$ walls. The α_p-chamber $p < n$ is obtained by repeatedly interchanging x_i and $x_{i+1} (i \neq p)$ and negating $x_i (i > p)$. Thus it is given by

$$\{x \in \mathbf{P} : \min_{i\leq p}(x_i) \geq \max_{i>p} |x_i|\}.$$

So things are identical to the previous case if $\alpha \neq \alpha_+, \alpha_-$. Adopt the same notation used for B_n and C_n. The fundamental α_p-cell is

$$\{x \in \mathbf{P} : rank'(x,1) \leq p, x_1 \geq 0\}.$$

Again if $\mathbb{W}$ is an α_p-chamber that does not lie in the fundamental cell then $\mathbb{W} = \mathbb{W}[-1|]$ or $\mathbb{W} = \mathbb{W}[|1]$. The arguments are now verbatim those of B_n and C_n except we need not assume that $p \leq n-2$ in the proof of fact 4.

LEMMA 5.2. *If $\alpha = \alpha_-$ (or α_+) then $L^- = \phi$.*

PROOF. We must analyze the α_+ and α_--cells. Rather than break the symmetry consider Q-chambers instead where $Q = \{\alpha_+, \alpha_-\}$. The Q-chamber containing W_+ is obtained by interchanging repeatedly x_i and x_{i+1} for $i \neq n$ so it is given by

$$\{x \in \mathbf{P} : \min_{i\leq n}(x_i) \geq |x_{n+1}|\}.$$

The smallest union of Q-chambers to contain the fundamental α_{n-1}-cell (called the fundamental Q-cell) is

$$\{x \in \mathbf{P} : rank'(x,1) \leq n, x_1 \geq 0\}.$$

Q-chambers are smaller than α_--chambers, so the fundamental α_--cell will contain the fundamental Q-cell. The notation introduced earlier for B_n and C_n extends to Q-chambers. In particular we have sets $L_Q = L_Q^+ \cup L_Q^-$.

Let $\mathbf{W}$ be a Q-chamber. If $\mathbf{W} = \mathbf{W}[1|]$, $\mathbf{W}$ is contained in the fundamental Q-cell; if $\mathbf{W} = \mathbf{W}[|1]$ then $\mathbf{W} \subseteq L_Q^+$; and if $\mathbf{W} = \mathbf{W}[-1|]$ then $\mathbf{W} \subseteq L_Q^-$. Let $\mathbb{W}$ be the α_--chamber containing $\mathbf{W}$. It is a union of Q-chambers. If $\mathbb{W}$ lies in the fundamental Q-cell then $\mathbb{W}$ must lie in the fundamental α_--cell. If $\mathbf{W} \subseteq L_Q^+$ ($\mathbf{W} = \mathbf{W}[x|1]$), then the wall $\mathbf{W}\langle x, 1\rangle$ of $\mathbf{W}$ is either of type α_+ placing $\mathbb{W}$ in the fundamental α_--cell, or of type α_- placing $\mathbb{W}$ in $K_n \cup L_n^+$ where K_n is the fundamental α_--cell. Finally if $\mathbf{W} \subseteq L_Q^-$ ($\mathbf{W} = \mathbf{W}[-1|x]$) then either the wall $\mathbf{W}\langle -1, x\rangle$ or the wall $\mathbf{W}\langle -1, -x\rangle$ is of type α_+. Passing through that wall one obtains another Q-chamber $\mathbf{W}'$ contained in $\mathbb{W}$ but one which is also in L_Q^+. $\mathbf{W}' \subseteq L_Q^+$ so $\mathbb{W} \subseteq K_n \cup L_n^+$. This proves the lemma.

6. A Lemma about Cells.

The subregular point p remains fixed through this section. The same assumptions on the Dynkin diagram remain in force, *i.e.* $\Delta = A_n, B_n, C_n$ or D_{n+1}. Assume that $|S| \geq 2$. Let γ_1 be the smallest root in S_- with the ordering on the roots given in section 5.

LEMMA 6.1. *Let $Y(\neq \mathbb{R}^n, \emptyset)$ be a union of γ_1-chambers such that if ξ is a γ_1-wall, $\xi \subseteq Y$, and $\xi \not\subseteq \overline{Y^c}$ then ξ vanishes. Suppose further that Y is path connected and every big wall of $Y \cap \overline{Y^c}$ is non-zero. Then $\overline{Y^c}$ contains a γ_1-cell.*

PROOF. If $\gamma_1 = \alpha_1$, a γ_1-cell is a γ_1-chamber, and $\overline{Y^c}$ is a union of γ_1-chambers so the lemma follows from $Y \neq \mathbb{R}^n$. Since $|S| > 1$ and γ_1 is the *smallest* in S_-, $\gamma_1 = \alpha_p$, $p \neq n$ (3.1) and since we are assuming $\Delta \neq F_4, G_2$ we are reduced to the case where α_p and α_{p-1} have the same length.

Let ξ_0 be a (non-zero) big wall of Y with $\xi_0 \subseteq \overline{Y^c}$ and let $\mathbb{W}_0$ be a big γ_1-chamber in Y which contains ξ_0. By (5.1), we have an γ_1-cell K which does not contain $\mathbb{W}_0$ but contains ξ_0. We shall show that K is the desired cell. We might as well translate the data by an element of the Weyl group and assume that K is the fundamental γ_1-cell. Let Y^0 be the path component of $Y \cap \overline{K^c}$ inside $\overline{K^c} = L$ containing $\mathbb{W}_0$. Y^0 is again the union of γ_1-chambers. It is enough to prove that the external walls in Y^0 are non-zero for then $Y = Y^0$. In terms of the following diagram with $Y^0 = A, Y = A \cup B \cup C$, $K = B \cup D \cup E$ we wish to show $B = C = D = \phi$ by proving the walls separating A and B are non-zero.

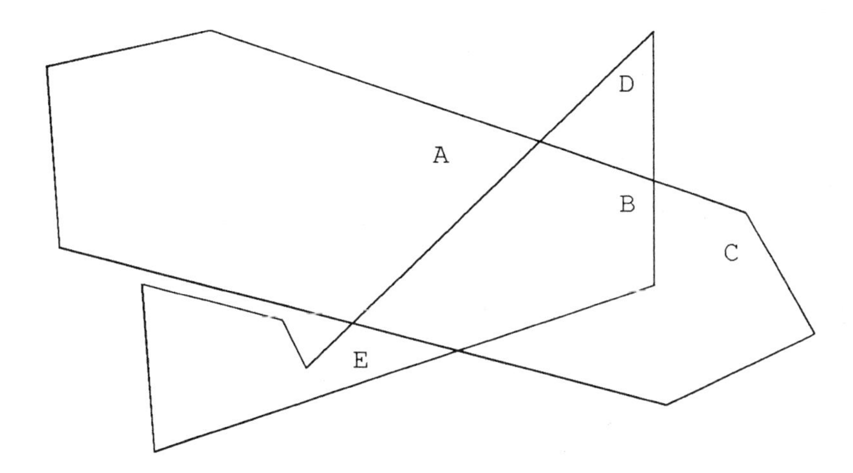

Let $\mathbb{W}$ be a big chamber in Y^0 containing a non-zero external wall $\xi \subseteq K$. If ξ' is also an external wall of $\mathbb{W}$ which is obtusely adjacent to ξ, then fact 2 shows that ξ and ξ' meet at a node A_2 of type (α_{p-1}, α_p). Since $\alpha_{p-1} \notin S \supseteq R$ then $z_1(W, \alpha_{p-1}) \neq 0$ for all W, and hence either all roots of type α_p at the node are zero or none are; so ξ' is a non-zero wall as well (consult the list of zero patterns). Fact 3 shows then that if ξ, ξ' are external walls in $\mathbb{W}$ then either both are zero or neither is.

Let $Y^0_+ = L^+ \cap Y^0$ and $Y^0_- = L^- \cap Y^0$. If two γ_1-chambers $\mathbb{W}$ and $\mathbb{W}'$ of Y^0_+ are adjacent and separated by ξ'', then by the hypothesis of the lemma, ξ'' is zero. Fact 4 gives walls ξ, ξ' of $\mathbb{W}$, $\mathbb{W}'$ resp. and a big node with angles $2\pi/3$. Since ξ'' is zero and the sum of the wall variables around a node is zero the external walls of $\mathbb{W}$ and $\mathbb{W}'$ vanish or do not vanish together.

The argument is nearly complete for A_n. Given $\mathbb{W}_0$ and any other γ_1-chamber $\mathbb{W}$ in $Y^0_+ = Y^0$ there is a path in Y^0_+ joining $\mathbb{W}$ and $\mathbb{W}'$ because Y_0 is defined as a path component. Since the external wall ξ_0 does not vanish repeated application of the previous paragraph shows that the external walls do not vanish. This shows that the external walls in Y^0 are non-zero and consequently that $Y = Y^0$.

From here on assume that the Dynkin diagram is of type B, C, or D, and that all chambers are γ_1-chambers unless specifically stated otherwise. Say that two γ_1-chambers $\mathbb{W}$ and $\mathbb{W}'$ are *proximate* if $\mathbb{W} = \mathbb{W}[i|]$ and $\mathbb{W}' = \mathbb{W}[-i|]$ for some i.

LEMMA 6.2. *If two γ_1-chambers $\mathbb{W}$, $\mathbb{W}'$ in Y^0 are proximate then the external walls in one are non-zero if and only if the external walls in the other are non-zero.*

PROOF. Let $\mathbb{W}$ and $\mathbb{W}'$ be given by $\mathbb{W} = \mathbb{W}[i|1]$ and $\mathbb{W}' = \mathbb{W}[-i|1]$. If $\mathbb{W}$ has a vanishing external wall then by the preceding arguments, the external wall $\mathbb{W}\langle i, 1\rangle$ vanishes. It follows then that $\mathbb{W}\ [1|i]$ must also lie in Y. Since $\mathbb{W}[1|i]$ and $\mathbb{W}[-i|1]$ both lie in Y the wall between them must be zero. This is an external wall, so $\mathbb{W}'$ contains a vanishing external wall.

To complete the proof of (6.1) for B, C, D it is enough to show that for any two big chambers $\mathbb{W}, \mathbb{W}'$ in Y^0_+ there exists a chain of chambers in Y^0_+ :

$\mathbb{W} = \mathbb{W}_1, \dots, \mathbb{W}_p = \mathbb{W}'$ such that $\mathbb{W}_i$ and $\mathbb{W}_{i+1}$ are adjacent or proximate $i = 1, \dots, p-1$. To do this first we pick any path from $\mathbb{W}$ to $\mathbb{W}'$ in the path connected space Y^0. If the path actually lies in Y^0_+ then we are done. Otherwise along the parts of the path which lie in Y^0_- we must construct a chain $\mathbb{W}_1, \dots, \mathbb{W}_p$ in Y^0_+ which runs alongside the portions of the path in Y^0_- such that $\mathbb{W}_i$ and $\mathbb{W}_{i+1}$ are adjacent or proximate.

Two adjacent chambers $\mathbb{W}$ and $\mathbb{W}'$ in Y^0_- are adjacent to a uniquely determined third chamber in Y^0_+. To see this set $\mathbb{W} = \mathbb{W}[-1x|y]$, $\mathbb{W}' = \mathbb{W}[-1y|x]$. Then the chamber $\mathbb{W}'' = \mathbb{W}[xy|1]$ lies in L^+. It also lies in Y^0 because the node $\mathbb{W}\langle x, y, 1\rangle$ is of type (α_{p-1}, α_p) and the wall $\mathbb{W}\langle x, y\rangle$ is zero. Thus $\mathbb{W}'' \subseteq Y_0^+$.

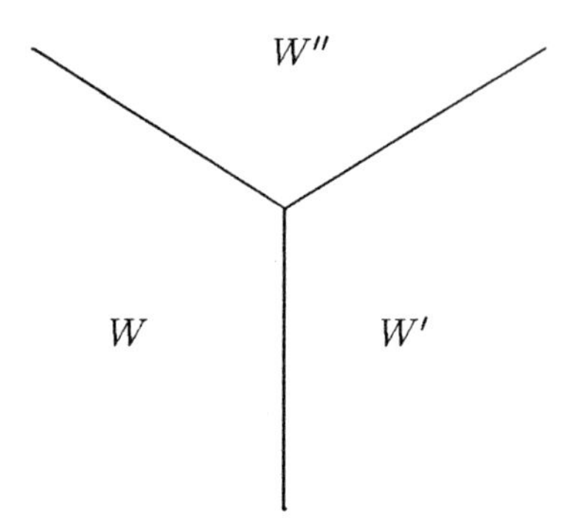

If we have three chambers $\mathbb{W}'_1, \mathbb{W}'_2, \mathbb{W}'_3$ in Y^0_- with $\mathbb{W}'_i, \mathbb{W}'_{i+1}$ adjacent $i = 1, 2$ then we have situation of the following diagram with $\mathbb{W}_1$ and $\mathbb{W}_2$ the chambers in Y^0_+ determined by the last paragraph.

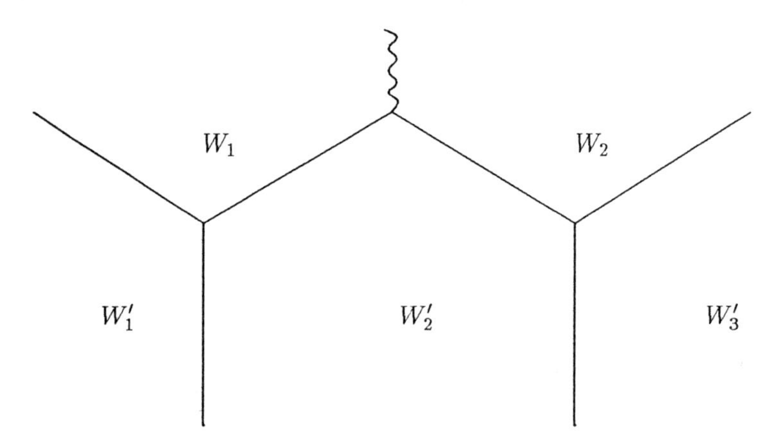

I claim that $\mathbb{W}_1$ and $\mathbb{W}_2$ are equal, adjacent, or proximate. Set

$$\mathbb{W}_1 = \mathbb{W}'_2[u|1] \qquad \mathbb{W}_2 = \mathbb{W}'_2[x|1]$$

$$\mathbb{W}'_1 = \mathbb{W}'_2[-1u|v] \qquad \mathbb{W}'_2 = \mathbb{W}'_2[-1|] \qquad \mathbb{W}'_3 = \mathbb{W}'_2[-1x|y].$$

So if $u \neq \pm x$ we have $\mathbb{W}_1 = \mathbb{W}_2[u|x]$ and they are adjacent; if $u = x$ they are equal, and if $u = -x$ they are proximate.

There is one last thing to check. If $\mathbb{W}_2 \subseteq Y_+^0$, if $\mathbb{W}_1'$, $\mathbb{W}_2' \subseteq Y_0^-$, and if $\mathbb{W}_1'$ and $\mathbb{W}_2'$ are adjacent as are $\mathbb{W}_2'$ and $\mathbb{W}_2$, then $\mathbb{W}_1$ and $\mathbb{W}_2$ are equal, adjacent, or proximate. The same proof applies leaving out $\mathbb{W}_3'$.

7. Contact.

In this section we show that the graph Γ' obtained in (4.2) is a subgraph of Γ_0. In other words we show that no vertices were eliminated in step 3 of the modification (4.2). It follows that if a linear subgraph in Γ_0 has extremal vertices which are extremal in S_- then the subgraph is S_-. To show that Γ' is a subgraph it is sufficient to show that if v_i and v_{i+1} are vertices in Γ' corresponding to adjacent roots α_i and α_{i+1} then there is a special node of type (α_i, α_{i+1}) in the modified star. This result clearly implies that v_i and v_{i+1} were already joined by an edge in Γ_0 and (since Γ_0 is a tree) that Γ' is a subgraph of Γ_0.

Continuing with the arguments of sections 5, 6, and 7 we could characterize the graphs Γ_0 associated with the zero patterns and classify the zero patterns according to the graphs. One can prove for instance for A_n that $\Gamma' = \Gamma_0$ so that Γ_0 is a subgraph of S and that Γ_0 determines the zero pattern up to Weyl group symmetry. We will not pursue these lines of enquiry further here.

Let β be the largest simple root in S_- (ordering the simple roots as in section 5). Let α be the next largest root in S_-. By (4.2) there is a union of α-chambers Z_α^- and a union of β-chambers Z_β such that the non-zero walls of type α are contained in Z_α^-, the non-zero walls of type β are contained in Z_β, Z_α^- is bounded by non-zero walls of type α, Z_β is bounded by non-zero walls of type β, and the interiors of Z_α^- and Z_β are disjoint. It follows from (6.1) and induction using the definition of cells that Z_α^- contains an α-cell K_α. By translating Z_α^-, Z_β by an element of the Weyl group we may assume that Z_α^- contains the fundamental α-cell. Let $L^- = L_\beta^-$ be the set defined in section 5 corresponding to the root β.

Similarly for every pair of adjacent roots $\alpha_i, \alpha_{i+1} \in S_-$ we can pick regions $Z_{\alpha_i}^-$ and $Z_{\alpha_{i+1}}$ such that the interiors of $Z_{\alpha_i}^-$ and $Z_{\alpha_{i+1}}$ are disjoint, the bounding walls are non-zero, and

$$Z_{\alpha_i}^- \subseteq Z_{\alpha_{i+1}}^-$$

$$Z_{\alpha_i} \supseteq Z_{\alpha_{i+1}}$$

LEMMA 7.1. *$Z_\beta \not\subseteq L_\beta^-$. If $\beta = \alpha_+ \in S_-$ (D_{n+1}) then $Z_\beta \not\subseteq L_Q^-$.*

PROOF. For A_n $L_\beta^- = \phi$ and the lemma is trivial. Let $Z = Z_\beta, L^- = L_\beta^-$. Assume that $\beta = \alpha_i$ (with the restrictions $i \neq n, n-1$ $B_n, C_n; \beta \neq \alpha_+, \alpha_-$ D_{n+1}). If $\mathbb{W} \subseteq Z \cap L^-$ then $\mathbb{W} = \mathbb{W}[-1|xy]$. The walls of type α_{i+1} of the (α_i, α_{i+1}) node $\mathbb{W}\langle -1, x, y\rangle$ are non-zero (α_{i+1} is not in S). By our restrictions it is a node of type A_2. So by the zero patterns for A_2 either the α_p-walls are zero which implies that $\mathbb{W}[x|1y]$ and $\mathbb{W}[y|1x]$ lie in Z_β or the α_p-walls are nonzero which implies that either $\mathbb{W}$ $[x|1y]$ or $\mathbb{W}[y|1x]$ lies in Z_β (for by assumption all non-zero walls of type α_p lie in Z_β).

Now assume that $i = n-1$, that the group is of type B_n or C_n and that $\mathbb{W} = \mathbb{W}[-1|x] \subseteq Z \cap L^-$. An (α_{n-1}, α_n) node is $\mathbb{W}\langle \pm 1, \pm x\rangle$.

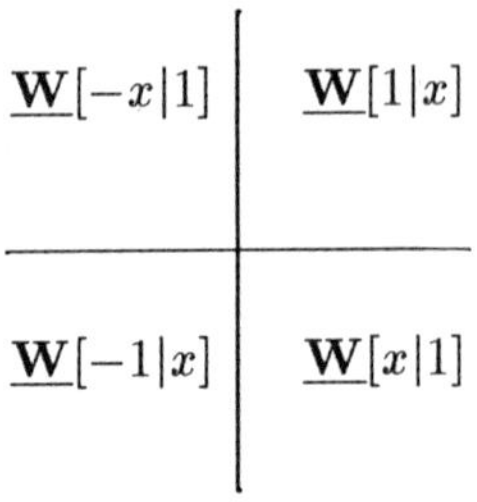

$\underline{\mathbf{W}}[-x\|1]$	$\underline{\mathbf{W}}[1\|x]$
$\underline{\mathbf{W}}[-1\|x]$	$\underline{\mathbf{W}}[x\|1]$

If a wall of $\mathbb{W}$ is zero then either $\mathbb{W}[-x|1]$ or $\mathbb{W}[x|1]$ is contained in $Z_\beta \cap L^+$. If the walls of $\mathbb{W}$ are non-zero then all the walls of type α_{n-1} at the node are non-zero. (Consult the list of zero patterns). The condition that Z_β contain all non-zero walls of type α_{n-1} forces $\mathbb{W}[-x|1]$, $\mathbb{W}[1|x]$, or $\mathbb{W}[x|1]$ to lie in Z. The chambers $\mathbb{W}[x|1]$ and $\mathbb{W}\ [-x|1]$ lie in L^+ and $\mathbb{W}[1|x]$ lies in K_β.

If $\beta = \alpha_n$ (type B_n, C_n) or $\beta = \alpha_+, \alpha_-(D_{n+1})$ then $L^- = \phi$ and there is nothing to prove. Turn to the second statement of the lemma. If $\mathbf{W} = \mathbf{W}[-1|x] \subseteq L_Q^-$ is a Q-chamber in Z_β then either $\mathbf{W}\langle -1, x\rangle$ or $\mathbf{W}\langle -1, -x\rangle$ is an α_--wall so that $\mathbf{W}[x|1]$ or $\mathbf{W}[-x|1]$ lies in $Z_\beta \cap L^+$.

LEMMA 7.2. *$Z_{\alpha_i} \not\subseteq L^-_{\alpha_i}$ for all $\alpha_i \in S$.*

PROOF. By the previous lemma we may assume that $\alpha_i \neq \beta$. Suppose that $\mathbb{W}[x|1]$ is an α_{i+1}-chamber which is contained in $Z_{\alpha_{i+1}}$. Then since $Z_{\alpha_i} \supseteq Z_{\alpha_{i+1}}$, $\mathbb{W}[|1x]$ must lie in Z_{α_i}. Similarly if $\mathbb{W} = \mathbb{W}[\pm 1|]$ is an α_{i+1}-chamber in $Z_{\alpha_{i+1}}$ then $\mathbb{W}[|1]$ lies in Z_{α_i}. Finally for D_{n+1}, $\alpha_i = \alpha_{n-1}$ we note the same arguments hold using Q-chambers instead of α-chambers.

LEMMA 7.3. *There is a special node of type (α_{i-1}, α_i) for $\alpha_{i-1}, \alpha_i \in S$.*

PROOF. First suppose the group is of type B_n, C_n, or D_{n+1} and $i \le n-1$. Set $Z = Z_{\alpha_i}$ and $Z^- = Z^-_{\alpha_{i-1}}$, and let K denote the fundamental α_{i-1}-cell. $K \subseteq Z^-$. Select $\mathbb{W} \subseteq Z$, $\mathbb{W} \not\subseteq L^-$. If $\mathbb{W} = \mathbb{W}\ [1x|y]$ then $\mathbb{W}' = \mathbb{W}[1|xy] \subseteq K$ and the interiors of $\mathbb{W}'$ and $\mathbb{W}$ are not disjoint (*contradiction*). So $\mathbb{W} = \mathbb{W}[xy|1]$. The node $\mathbb{W}\langle 1, x, y\rangle$ contains the α_{i-1}-chamber $\mathbb{W}[1|xy] \subseteq K$. This is the special node.

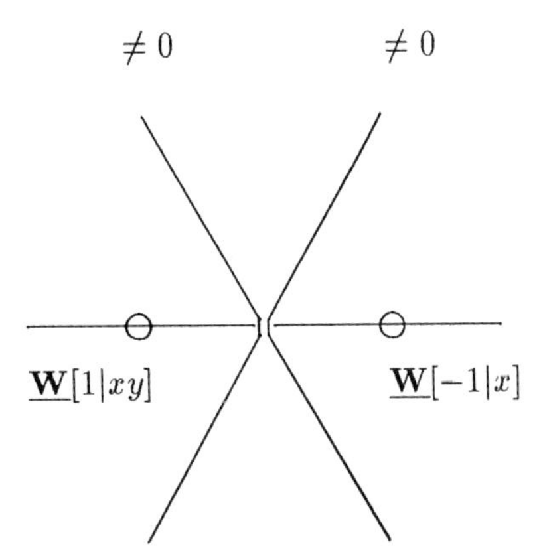

Now assume $i = n$, and the group is of type B_n or C_n. If $\mathbb{W} = \mathbb{W}[1x|]$ then the α-chamber $\mathbb{W}' = \mathbb{W}[1|x]$ lies in K and the interiors of $\mathbb{W}'$ and $\mathbb{W}$ are not disjoint (*contradiction*). So $\mathbb{W} = \mathbb{W}\,[-1x|]$. The node $\mathbb{W}\langle\pm 1, \pm x\rangle$ contains the α_{i-1}-chamber $\mathbb{W}[1|-x]$. This is a special node.

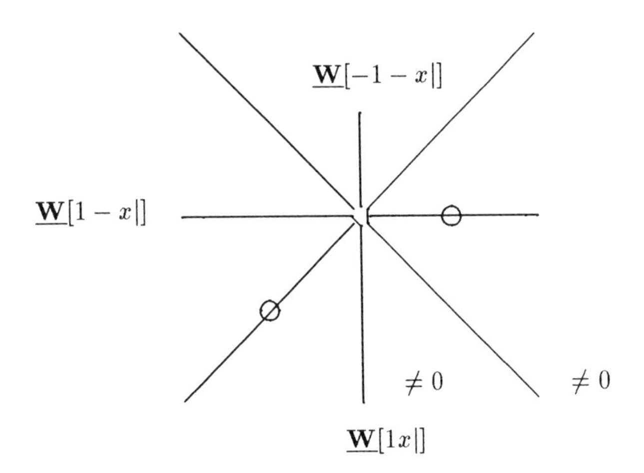

Now assume that the group is of type D_{n+1} and $\alpha_i = \alpha_-$, $\alpha_{i-1} = \alpha_n$. Let $\mathbf{W}$ be a Q-chamber in Z not in L_Q^-. If $\mathbf{W} = \mathbf{W}[1x|y]$ then the α_{i-1}-chamber $\mathbf{W}[1|xy]$ lies in K hence in Z^- and intersects the interior of $\mathbf{W}$ (contradiction). So $\mathbf{W} = \mathbf{W}[xy|1]$. The node $\mathbf{W}\langle 1, x, y\rangle$ is of type $(\alpha_{n-1}, \alpha_\pm)$. The α_{n-1}-chamber $\mathbf{W}[1|xy]$ lies in $K \subseteq Z^-$. Since Z^- and Z are disjoint and bounded by nonzero walls this must be a special node.

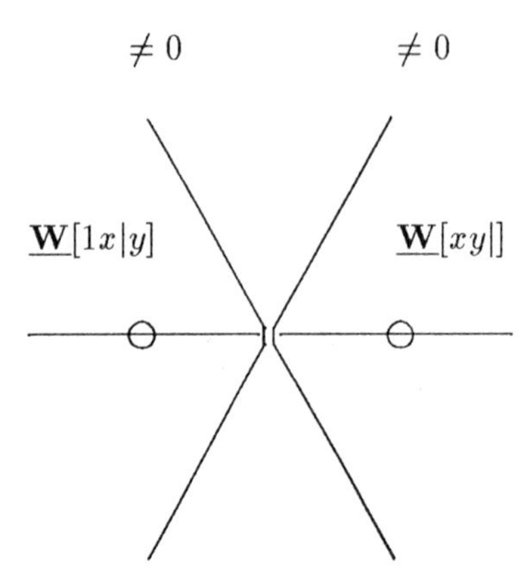

For A_n we make use of graph automorphism and the fact that an α_n-cell is precisely the closure of the complement of an α_n-chamber. Let a fundamental *dual* α_i-cell be the smallest union of α_i-cells to contain a fundamental dual α_{i+1}-cell and let a fundamental dual α_n-cell be the fundamental α_n-chamber. It follows by induction that a dual α_i-cell is the closure of the complement of an α_i-cell. Z must contain a dual α_i-cell and Z^- contains an α_{i-1}-cell. Since the interiors of Z and Z^- are disjoint Z is equal to the dual cell, and Z^- is equal to K. The result follows.

8. The Assumption 3.1.

Assumption 3.1 was made to simplify the arguments in section 3. The cases where the assumption fails to hold are easily managed. We have for instance:

LEMMA 8.1. *If the roots of G are all the same length and $\theta_1(W_0, \alpha_0) = 0$ for some (W_0, α_0) then assumption 3.1 holds.*

PROOF. The hypothesis $\theta_1(W_0, \alpha_0) = 0$ of the lemma is made to insure that p does not lie in Y''. First we show that $|S| \geq 2$. Suppose that $|S| = 1$. Form a path $W_0, \ldots, W_p$ from W_0 to W_p where W_p is chosen so that $\theta_1(W_p, \alpha_0) \neq 0$. Let i be the smallest index for which $\theta_1(W_i, \alpha_0) \neq 0$. Then $i > 0$ and the wall dividing W_{i-1} from W_i is not of type α_0. Say it is of type β_1. It is easy to see that $(\alpha_0, \beta_1) \neq 0$. The assumption on the lengths of roots forces it to be a node of type A_2. Since $\beta_1 \notin S$ ($|S| = 1$) the walls of type β_1 at the node are non-zero. There is a non-zero wall of type α_0 as well as a vanishing wall of type α_0 at the node. But this is impossible by the zero pattern $(A.I)$ of type A_2. Thus $|S| \geq 2$.

For any root $\beta \in S, \tilde{z}(W, \beta)x(W, \beta) = 0$ (1.2). Lemma 3.2 holds because it only makes use of the assumption $|S| \geq 2$. Fix (W_2, β_2) with $\theta_1(W_2, \beta_2) \neq 0$. Then $x(W_2, \beta_2) = 0$. For any chamber W_3 adjacent to W_2, $x(W_3, \beta_2) = 0$ except if the wall separating W_2 and W_3 has type β_3 for some $\beta_3 \in R$ and $\theta_1(W_2, \beta_3) \neq 0$. Then $x(W_3, \beta_3) = 0, x(W_2, \beta_3) = x(W_2, \beta_2) = 0$, and $(\beta_2, \beta_3) \neq 0$ (1.3). It follows by considering paths originating from W_2 that R forms a connected Dynkin diagram and for every Weyl chamber W there is a simple root $\beta \in R$ such that $x(W, \beta) = 0$. Also if $\alpha \in S$ then $x(W, \alpha) = 0$ for some W and $(\alpha, \beta) \neq 0$ for some $\beta \in R$. It follows that if $R \neq \phi$ then for $\beta_1, \beta_2 \in S$ and $(\beta_1, \beta_2) \neq 0$

either β_1 or β_2 lies in R. Say $\beta_2 \in R$. Then select W_2 such that $\theta_1(W_2, \beta_2) \neq 0$ then $B(W_2) \neq B(W_3)$ where W_3 is the Weyl chamber through a wall of type β_2 from W_2. So either $x(W_2, \beta_1) \neq 0$ or $x(W_3, \beta_1) \neq 0$ (the line of type β_2 does not intersect two lines of type β_1). Therefore the walls (W_2, β_1) and (W_2, β_2) do not come together at a solid node.

Finally we consider the case $R = \phi$. Then $B(W) = B(W')$ for all W and $x(W, \beta) = x(W', \beta)$ for all W, W', β. If $\alpha \in S$ then $x(W, \alpha) = 0$ for some W so that $x(W, \alpha) = 0$ for all W. This forces $|S| = 2, S = \{\alpha, \beta\}$ with $(\alpha, \beta) \neq 0$. Suppose for a contradiction that there is a solid node of type (α, β). Let W be a chamber at the node. If we reflect through the wall (W, γ) with $(\gamma, \alpha) = (\gamma, \beta) = 0$ then by the nature of the nodes of type $A_1 \times A_1$ the solid node is reflected to another solid node:

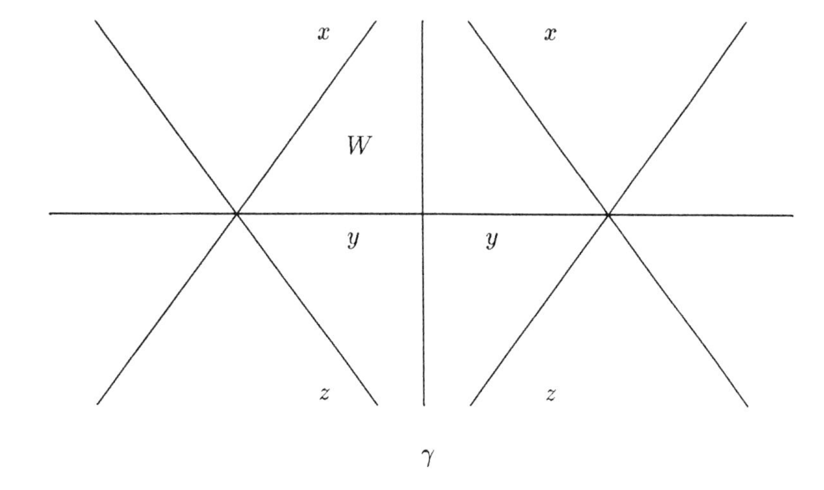

If we reflect through the wall (W, γ) with $(\gamma, \alpha) = 0$, $(\gamma, \beta) \neq 0$ then by the nature of the nodes of type A_2 the solid node is reflected to another solid node $(\theta_1(W', \gamma) \neq 0$ for all $W')$:

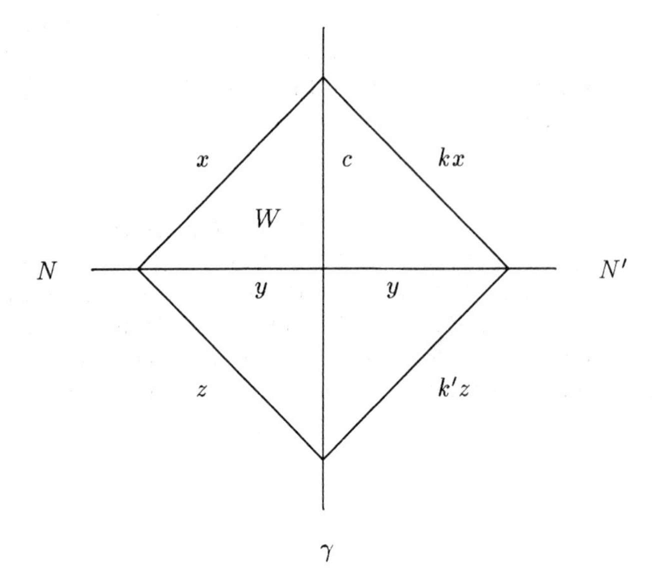

By repeatedly reflecting through walls we find that every node of type (α, β) is a solid node. This contradicts the assumption that $\alpha, \beta \in S$.

REMARK 8.2. The same conclusion holds under the weaker assumption that all the roots in S or adjacent to a root in S are the same length. The assumption $|S| = 1$ is made to avoid trivialities. There are a few cases where assumption 3.1 breaks down; but they too are easily classified. For example one can have a solid node of type (α, β) in the situation $S = \{\alpha, \beta\}$ where α and β have different lengths. I will not treat these cases here.

V. THE SUBREGULAR FUNDAMENTAL DIVISOR

This chapter and the next develop an integral expression for the subregular germ of a κ-orbital integral of a Cartan subgroup T. The data entering into this expression are a surface (together with a description of its irreducible components and F-structure), a character $\kappa(E)$ of $F^{\times}$ for each irreducible component of the surface defined over F, a properly normalized 2-form on each irreducible component of the surface, a cocycle depending on the F-rational points of the surface of $\mathrm{Gal}(\bar{F}/F)$ with values in T, a character κ on $H^1(T)$, and finally canonical coordinates (w, ξ) on the surface. The germ is obtained by evaluating the cocycle by κ and integrating it with respect to the 2-form over the surface. We will see that the principal value integrals cannot be given an intrinsic coordinate free definition. Canonical coordinates are introduced to overcome this shortcoming.

Section 1 shows that the irreducible components of the surface are in bijection with the lines of the Dynkin curve and that each component is a rational surface. In section 2 we give a condition that must be satisfied for the 2-forms on the various irreducible components to be compatibly normalized. Near F-rational points at the intersection of two irreducible components of the surface the definition of the principal value integral differs from the usual one. In particular, the definition is not coordinate free. This places certain restrictions on coordinates near F-rational points at the intersection. These restrictions are discussed in section 3. Section 4 develops formulas giving the contribution to the subregular germ of the divisors whose intersection is defined over F, which are not themselves defined over F. These formulas will be applied to the subregular germ of $G = {}^2A_{2n}$ in chapter VII. Section 5 gives formulas for coordinate transitions on the overlap of two patches. Section 6 develops some useful coordinate relations that will be used to study the rationality of the surface in chapter VI.

1. Regularity.

Let Y_s be the open subvariety of Y'' of elements $(b, B(W))$ such that b is regular or subregular. (b need not be *unipotent*.) Let π denote the restriction to Y_s of the morphism from Y'' to G described in chapter I.

LEMMA 1.1. *On Y_s*

a) *If $\lambda = 0$ and $\pi(p) = u$ is subregular then either*

   i) *there exists a simple root α' such that $x(\alpha) \neq 0$ for $\alpha \neq \alpha'$ and $x(\alpha') = 0$, or*

   ii) *there exist adjacent simple roots α', α'' such that $x(\alpha) \neq 0$ for $\alpha \neq \alpha', \alpha''$, and $x(\alpha') = x(\alpha'') = 0, x(\alpha' + \alpha'') \neq 0$.*

b) *Y_s is regular and the divisors of $\lambda = 0$ have normal crossings.*

c) *The divisors on Y_s are E_α $\forall$ α, and E_0 the regular divisor.*

d) $E_\alpha \cap E_{\alpha'}$ *is non-empty if and only if* $(\alpha, \alpha') \neq 0$.

REMARK. By the coordinate free description of E_Σ given in $(II.9.3)$, it follows that $E_\Sigma \neq E_{\Sigma'}$ if $\Sigma \neq \Sigma'$. Here E_α denotes E_Σ, $\Sigma = \{\alpha\}$, and $E_0 = E_\Sigma$, $\Sigma = \emptyset$.

PROOF. (a) Option (i) corresponds to the situation where $B(W_+)$ lies in only one line, a line of type α'. If $B(W_+)$ lies in two lines of $(B\backslash G)_u$ then the lines must correspond to adjacent roots α' and α''. Suppose that $x(\alpha') = x(\alpha'') = x(\alpha' + \alpha'') = 0$. We show that this is not subregular. If α' is at least as long as α'' then $\sigma_{\alpha'}(\alpha' + \alpha'') = \alpha''$. It follows that if u is unipotent and $x(\alpha') = x(\alpha'') = x(\alpha' + \alpha'') = 0$, then $u^{\exp(x_1 X_{-\alpha'})} \in N_{\alpha''}$ and

$$u^{\exp(x_1 X_{-\alpha'})\exp(x_2 X_{-\alpha''})} \in N_{\alpha''} \subseteq B_0$$

or

$$u \in B_0^{\exp(-x_2 X_{-\alpha''})\exp(-x_1 X_{-\alpha'})}.$$

This shows that $\dim(B\backslash G)_u \geq 2$ so that u is not subregular. This proves (a).

We prove $(b), (c)$, and (d) together. Case 1. Suppose at $p, x(\alpha) \neq 0$ for $\alpha \neq \alpha', \alpha$ simple. To prove regularity we show that the local ring is generated by $x(\gamma) : \gamma$ positive and $z(\alpha')$. The equations $(II.4.2)$

$$w(\alpha) = 1 : \alpha \text{ simple.}$$

$$z(\alpha) = \lambda / x(\alpha) : \alpha \neq \alpha'$$

$$w(\gamma) = z(\gamma - \alpha) x(\gamma)/x(\alpha) : \gamma \text{ not simple}$$

$$\text{where } z(\beta) \text{ is defined to be } \prod z(\alpha)^{m(\alpha)} \text{ for } \beta = \sum m(\alpha)\alpha.$$

$$\lambda = x(\alpha') z(\alpha') : \lambda$$

show that $z(\alpha)\ \forall\ \alpha, w(\gamma) : \gamma$ positive and λ lie in the ring generated by $x(\gamma) : \gamma$ positive and $z(\alpha')$. This proves regularity. The equation $\lambda = x(\alpha')z(\alpha')$ shows that there are two divisors on this coordinate patch and that they have normal crossings. The divisor $E_{\alpha'}$ is defined by $x(\alpha') = 0$ and the divisor E_0 is defined by $z(\alpha') = 0$.

Case 2. Suppose at p, $x(\alpha) \neq 0$ for $\alpha \neq \alpha', \alpha''$, and $x(\alpha' + \alpha'') \neq 0$. Then by $(II.4.2)$

$$z(\alpha) = \lambda / x(\alpha) : \alpha \neq \alpha', \alpha''$$

$w(\gamma) = z(\gamma - \alpha) x(\gamma)/x(\alpha)$: for γ simple and γ not in the rank two system generated by α', α''

$$z(\alpha') = w(\alpha' + \alpha'') x(\alpha'')/x(\alpha' + \alpha'')$$

$$z(\alpha'') = w(\alpha' + \alpha'') x(\alpha')/x(\alpha' + \alpha'')$$

$$\lambda = x(\alpha') x(\alpha'') w(\alpha' + \alpha'')/x(\alpha' + \alpha'')$$

$$w(\delta) = x(\delta) w(\gamma) z(\delta - \gamma)/x(\gamma) : \text{ for } \delta = m(\alpha')\alpha' + m(\alpha'')\alpha''$$

where $\gamma = \alpha' + \alpha''$.

$x(\alpha') = 0$ defines $E_{\alpha'}$, $x(\alpha'') = 0$ defines $E_{\alpha''}$, $w(\alpha' + \alpha'') = 0$ defines E_0 and the local ring at p is regular for it is generated by $x(\delta)$: all δ and $w(\alpha' + \alpha'')$. These divisors have normal crossings. The only two subregular divisors on this patch are $E_{\alpha'}$ and $E_{\alpha''}$ and $(\alpha', \alpha'') \neq 0$. By part ($a$) of the lemma, patches of the form considered in case 1 and case 2 cover the subregular divisors. This completes the proof.

The patch described in case two of the lemma will be used frequently in this chapter. We denote it by $U(\alpha', \alpha'')$. It depends on the choice of opposite Borel subgroups (B_∞, B_0).

Fix a subregular element $u \in G$. Set $E_\alpha(u) = \pi^{-1}(u) \cap E_\alpha$ and let $E_\alpha(u)^0$ be an irreducible component of $E_\alpha(u)$.

LEMMA 1.2.

a) *Suppose that* $(u_1, (B_0^{n_w}))^\nu$ *and* $(u_2, (B_0^{n'_w}))^{\nu'}$ *both lie in* $E_\alpha(u)^0$. *Then* $\nu\nu'^{-1} \in P_\alpha$ *the parabolic of type* α *containing* B_0.

b) $E_\alpha(u)$ *is a disjoint union of* $|\alpha|^2/|\alpha_{min}|^2$ *rational surfaces where* α_{min} *is any short root.*

PROOF. We begin with (b). If

$$(u, (B(W))) \in E_\alpha(u) \text{ then } B(W_+) \in (B\backslash G)_u.$$

The equation (IV.1.2) $\lambda T(W, \alpha) = x(W, \alpha)z(W, \alpha)$ and $z(\alpha) \neq 0$ imply that $x(W, \alpha) = 0$ for all W. Thus $B(W)$ lies in a line of type α for all W.

n_w has the form:

$$n_w = \exp(z(W_q, \alpha_q)X_{-q}) \ldots \exp(z(W_1, \alpha_1)X_{-1})$$

where $W_+ = W_1, \ldots, W_q, W_{q+1} = W$ is a path from W_+ to W and W_{i+1} and W_i are separated by a wall of type α_i. From the relations $z(\alpha') = 0$ $\alpha' \neq \alpha$ it follows that $z(W, \alpha') = 0$ on E_α for all (W, α') $\alpha' \neq \alpha$. Thus the product for n_w collapses into an expression $n_w = \exp(a_w X_{-\alpha})$ on E_α where $a_w = \sum z(W_i, \alpha_i)$ and the sum is over all i such that $\alpha_i = \alpha$.

Write

$$(u, (B(W))) = (u^{\nu^{-1}}, (B_0^{n_w}))^\nu \in E_\alpha(u), \quad n_w = \exp(a_w X_{-\alpha}).$$

If $B(W_+) \in \ell_\alpha (= B_0\backslash P_\alpha \nu)$ then for all W, $B(W) = B_0^{n_w \nu} \in P_\alpha^\nu$ also lies in the line ℓ_α ($= B_0\backslash P_\alpha \nu$). Thus $E_\alpha(u)$ breaks up into a disjoint union of varieties according to which line of type α the $(B(W))$ lie in.

By [22,p.146], $C_G(u)$ acts transitively on the lines of a given type in $(B\backslash G)_u$. Thus there is a variety for each line of type α and they are isomorphic over $\bar{F}$. It remains to be seen that each is a rational surface.

Now turn to (a) B_0^ν and $B_0^{\nu'}$ lie in the same line ℓ_α, i.e. the same parabolic subgroup which must be $P_\alpha^\nu = P_\alpha^{\nu'}$. Thus $\nu\nu'^{-1} \in P_\alpha$.

We return to (b). Fix the component $E_\alpha(u)^0$ with associated line ℓ_α. We will often denote this component by $E(\ell_\alpha, u)$. Select B_0 to lie in two distinct lines of $(B\backslash G)_u$ including the line ℓ_α. Select a point $(u_1,(B(W)))$ in $E_\alpha(u)$ such that $B(W_+) = B_0$. Then $(u_1,(B_0^{n_w}))^\nu = (u_1,(B(W))) = (u_1,(B_0^{n_w}))$. So $u = u_1$ and $\nu = 1$. Let $(u_2,(B_0^{n'_w}))^{\nu'}$ be any other point in $E_\alpha(u)$. Then by (a), $\nu\nu'^{-1} = \nu'^{-1} \in P_\alpha$ so $\nu' = \exp(\xi X_{-\alpha})$ for some ξ. $u_2^{\nu'} = u$ or $u_2 = u^{\nu'^{-1}}$. It follows that for fixed u, the coefficients of u_2 are polynomials in ξ.

(1.1) shows that on an open set $U(\alpha,\alpha')$ (defined by $x(\alpha'') \neq 0$ $\alpha'' \neq \alpha,\alpha'$ $x(\alpha+\alpha') \neq 0$) the coefficients $x(\gamma)$ $\forall$ γ and $w(\alpha+\alpha')$ generate the local ring. Thus on an open patch for a fixed u, ξ and $w(\alpha+\alpha')$ are coordinates on $E_\alpha(u)^0$. This shows that $E_\alpha(u)^0$ is a rational surface.

2. Igusa theory and measures.

Each divisor E_α projects to G. The image of an open set in E_α is the set of subregular unipotent elements in G. We use this fibration to express the principal value integral over E_α as a repeated integral over the fibre of an element u and an invariant integral over the conjugacy class of u. We have seen that ξ and $w = w(\alpha+\beta)$ serve as coordinates on $U(\alpha,\beta) \cap E_\alpha(u)^0$. We compute the differential form on the fibre over u in this section.

It is often inconvenient to work directly with F-coordinates on the variety. We translate the formulas in [15] for the differential form ω_E on a divisor E into a more convenient form. It is not our purpose here to develop the formalism of Igusa theory. Rather we adapt the formalism to our specific purposes. For details see [11] or [15]. From [15] we have formulas

$$f = \gamma\kappa_1(\mu_1)\dots\kappa_n(\mu_n)$$

$$h(0,\mu_2,\dots,\mu_n) = \gamma\theta(\alpha)^{-1}\prod_{j=2}^{n}\kappa_j(\mu_j)\theta(\mu_j^{-a_j}) \text{ where } \theta^{a_1} = \kappa_1$$

$$\omega = W(\mu_1,\dots,\mu_n)\mu_1^{b_1-1}\dots\mu_n^{b_n-1}d\mu_1\dots d\mu_n$$

$$\omega_E = W(0,\mu_2,\dots,\mu_n)\alpha^{-\beta}\prod_{j=2}^{n}\mu_j^{b_j-\beta a_j-1}d\mu_2\dots d\mu_n$$

$$\lambda = \alpha\mu_1^{a_1}\dots\mu_n^{a_n}.$$

By these formulas we see that ω_E is the restriction of $\omega/(\lambda^\beta(d\mu_1/\mu_1))$ to E. The coordinate μ_1 may be replaced by any other coordinate μ' (not necessarily F-rational) such that $\mu' = 0$ defines E. We let $\mu' = x(\alpha)$ so that ω_E for $E = E_\alpha$ becomes the restriction (of the extension) of $\omega/(\lambda^\beta(dx(\alpha)/x(\alpha))$ to E. We obtain h by extending $f/\theta(\lambda)$ to E. Finally we note that $\kappa(E)$ can be described as the character such that $f/\kappa(E)(\mu)$ extends to an open set on E where as usual μ is a local coordinate and $\mu = 0$ defines E locally.

By remark $I.6.2$ the fibres $E(u)$ and $E(u')$ are isomorphic (over $\bar{F}$) by the G-action on Y_Γ. It is therefore sufficient to fix the measure for one unipotent element. The choice of isomorphism is not uniquely determined but it follows from the G-invariance of ω_Y that the identification of the form on $E(u)$ with one on $E(u')$ is independent of the choice of isomorphism. We fix a subregular element u independent of T and Γ. We may take ω_Y to be

$$\omega_Y = d\lambda \wedge dx_1 \wedge \ldots \wedge dx_p \wedge d\nu_1 \wedge \ldots \wedge d\nu_p.$$

By fixing isomorphisms over $\bar{F}$ between the varieties X_1 constructed for various Cartan subgroups T, we fix the form for all T.

We fix a 2-form on $E(u) \overset{(def)}{=} \bigcup E(\ell_\alpha, u)$ in two steps.

1) If ℓ_α and ℓ_β intersect we link the normalization of the 2-form on $E(\ell_\alpha, u)$ relative to that on $E(\ell_\beta, u)$ by matching their residues on $E(\ell_\alpha, u) \cap E(\ell_\beta, u)$. The following equalities hold on $E_\alpha \cap E_\beta$:

EQUATION 2.1.

$$\omega_Y x(\alpha)x(\beta)/(\lambda^2 dx(\alpha) \wedge dx(\beta)) = \omega_{E_\alpha} x(\beta)/dx(\beta) = \omega_{E_\beta} x(\alpha)/dx(\alpha).$$

Of course any other local coordinates may be substituted for $x(\alpha)$ and $x(\beta)$. This last equality determines the relative normalization on adjacent components.

2) We single out any component $E(\ell_\alpha, u)$ and fix a normalization of the measure. Any normalization is acceptable, but it must be independent of T and Γ.

REMARK 2.2. We give local coordinates on the subregular class O. Fix a parabolic subgroup P_α which is defined over F. Let B be a Borel subgroup in P_α. Consider the subset O' of O on which there is a line of type α fixed by $\mathrm{Gal}(\bar{F}/F)$ in $(B\backslash G)_y$ meeting $B^{N_{-\alpha}}$ for $y \in O'$. Then B^ν for some $\nu \in N_{-\alpha}(\bar{F})$ is a Borel subgroup in a line of type α in $(B\backslash G)_y$. We may assume that the line is given by P_α^ν. Since the line is defined over F, $\nu \in N_\alpha(F)$, and if $y \in G(F)$ then $y^{\nu^{-1}} \in N_\alpha(F)$. We may use the coefficients of $N_{-\alpha}(F)$ and $N_\alpha(F)$ as local coordinates on $O(F)$ in some neighborhood of y. The morphism $N_\alpha \times N_{-\alpha} \rightarrow O$ is defined over F. Note that the number of points in $N_\alpha(F) \times N_{-\alpha}(F)$ covering a given unipotent element y depends on the number of lines of type α in $(B\backslash G)_y$ meeting $B^{N_{-\alpha}}$.

LEMMA 2.3. *The form on a fibre $E(\ell_\alpha, u)$ is given up to a scalar by $\delta^{-1} d\xi \wedge dw/(\xi w^2)$ where $\delta = x(\beta)/(\xi x(\gamma))$. $(\gamma = \alpha + \beta)$.*

PROOF. By (2.2) we may use the coefficients of $N_\alpha(F)$ and $N_{-\alpha}(F)$ as local p-adic coordinates on $O(F)$.

Choose B_0 to lie in ℓ_α. Elements on an open set near E_α can be written in the form

$$(b, (B_0^{n_w}))^{\exp(\xi X_{-\alpha})^\alpha \nu}.$$

Now

$$\lambda = x(\alpha)x(\beta)w/x(\alpha+\beta)$$

and by the previous paragraph the restriction of the form to E_α is given by $\omega_E = (\omega_Y/\lambda^2)(x(\alpha)/dx(\alpha))$. We have $\omega_Y/\lambda^2 =$

$$x(\gamma)(dw/w^2) \wedge (dx(\alpha)/x(\alpha)) \wedge (dx(\beta)/x(\beta)) \wedge dx(\gamma) \wedge \prod dx(\alpha') \wedge (d\xi \wedge d^\alpha \nu).$$

$$\omega_E = x(\gamma)(dw/w^2) \wedge (dx(\beta)/x(\beta)) \wedge dx(\gamma) \wedge \prod dx(\alpha') \wedge (d\xi \wedge d^\alpha \nu).$$

The coefficients of b are $x(\beta), x(\gamma)$ etc. We can write $u = b^{\exp(\xi X_{-\alpha})}$, and write the coefficients of u as $u(\beta), u(\gamma)$, etc. Then the coefficients of u and the coefficients of ${}^\alpha\nu$ are local coordinates along the subregular conjugacy class. The relation $u = b^{\exp(\xi X_{-\alpha})}$ implies that

$$d\xi \wedge dx(\beta) \wedge dx(\gamma) \wedge \ldots = d\xi \wedge du(\beta) \wedge du(\gamma) \wedge \ldots$$

Thus when expressed in terms of the variables $u(\eta)$ instead of the variables $x(\eta)$, the form becomes

$$\omega_E = (x(\gamma)/x(\beta))(dw/w^2) \wedge du(\beta) \wedge du(\gamma) \wedge \prod du(\alpha') \wedge (d\xi \wedge d^\alpha \nu)$$

or

$$\omega_E = (x(\gamma)/x(\beta))(dw/w^2) d\xi \wedge \omega_{sub}$$

where ω_{sub} is independent of the coordinates on the fibre.

The tangent direction X of the regular curve Γ can be identified with a vector in the regular elements of the Lie algebra $Lie(T)$ of T. In the remainder of this text X will denote an element of $Lie(T)$. The expression for the germs will depend on the tangent direction X through the parameters $\alpha(X) : \alpha$ simple. To make this dependence explicit we introduce the field K_X of rational functions on $Lie(T)$. It is isomorphic to $\bar{F}(x_1, \ldots, x_n)$ where $n = \dim(T)$ and $x_1, \ldots, x_n$ are independent. We may identify the simple roots with elements of this field. The regular function $T(W, \alpha)$ discussed in $(IV.1.1)$ equals $\pm\eta(X)$ when $\lambda = 0$ where η is the root determined by the wall (W, α) of W.

For points of $E(\ell_\alpha, u)$ we have defined $\delta(\xi)$ by $\xi x(\gamma)/x(\beta) = \delta(\xi)^{-1}$. There is the simple but useful relation.

LEMMA 2.4. *On* $E(\ell_\alpha, u), z(W_+, \alpha)/\xi = \delta(\xi) w(\alpha + \beta)\alpha(X)$.

PROOF. By $(II.4.2)$,

$$\lambda = x(\alpha)x(\beta)w(\alpha+\beta)/x(\gamma) = (1 - \alpha^{-1}(t))x(\beta)w(\alpha+\beta)/(x(\gamma)z(W_+, \alpha)).$$

$$z(W_+, \alpha)/\xi = (x(\beta)/\xi x(\gamma))w(\alpha+\beta)(1-\alpha^{-1})/\lambda.$$

On $E(\ell_\alpha, u), (1-\alpha^{-1})/\lambda = \alpha(X)$ and $x(\beta)/\xi x(\gamma) = \delta(\xi)$.

3. Principal value integrals at points of $E_\alpha \cap E_\beta$.

If we wish to compute the principal value integral at a point near the intersection of two divisors E_α and E_β where $\kappa(E_\alpha) = \kappa(E_\beta)$, $a(E_\alpha) = a(E_\beta)$ and $b(E_\alpha) = b(E_\beta)$ then we must use the formulas for principal value integrals of [15].

We follow [15,p.469]. The Igusa constants $a(E_\alpha)$ and $a(E_\beta)$ are 1. The principal part of $\prod_{i=1}^{2}(1-t^{a_i})^{-1}$ at $t = 1$ is $\sum_{j=1}^{2} c_j(1-t)^{-j}$ with $c_1 = 0, c_2 = 1$. The polynomial $A(x)$ defined to be $\sum c_j(x+1)\ldots(x+j-1)$ is $(x+1)$. The polynomials $A_r(y)$ defined by $A(x-y) = A_1(y) + xA_2(y)$ are $A_2(y) = 1, A_1(y) = 1-y$. It follows from [15,p.470] that the contribution to the term $F_1(\beta, \theta, f), \beta = 2$ on a small patch with coordinates satisfying conditions the local conditions of [15] is given by

$$\int A_1(M)h_2|\nu_2|.$$

The integral extends over $U \cap E_\alpha \cap E_\beta$ where U is our coordinate patch. By the remarks of section 2 combined with the formulas in [15] we see that h_2 is given by $f/\theta(\lambda)|_{E_\alpha \cap E_\beta}$ and ν_2 is given by

$$(\omega/\lambda^\beta)x(\alpha)x(\beta)/(dx(\alpha)dx(\beta))|_{E_\alpha \cap E_\beta}.$$

We must still define M. We may assume that we are on a coordinate chart such that

$$\lambda = \alpha\mu_1\mu_2\mu_3^a$$

$$|\mu_i| \le q^{-m_i} \qquad i = 1, \ldots, n$$

$\mu_1 = 0$ defines E_α locally and $\mu_2 = 0$ defines E_β locally. $\mu_3 = 0$ defines E_0 locally if E_0 intersects the coordinate patch. If E_0 does not intersect the coordinate patch then $a = 0$. Then M is given by $m + m_1 + m_2 + am(\mu_3)$ where $q^{-m} = |\alpha|$ and $m(\mu_3) = -log_q|\mu_3|$ on our patch. We may restate this definition of M in terms of coordinates on E_α and E_β.

$(x_1, y_1) = (\mu_2, \mu_3)$ are local p-adic coordinates on a patch

$$U_1 = \{(x_1, y_1) : |x_1| \le q^{-m_1}, |y_1| \le q^{-n_1}\}$$

of E_α and $(x_2, y_2) = (\mu_1, \mu_3)$ are local p-adic coordinates on a patch

$$U_2 = \{(x_2, y_2) : |x_2| \le q^{-m_2}, |y_2| \le q^{-n_2}\}$$

of E_β. We have

A.i) $y_1 = y_2$ on $U_1 \cap U_2$

A.ii) $n_1 = n_2$

A.iii) If E_0 intersects U_1 or U_2 then $x_1 = 0$ defines E_0 in U_1 and $x_2 = 0$ defines E_0 in U_2.

A.iv) $|\alpha| = q^{-m}$ on $U_1 \cap U_2$.

We see that $M = m + m_1 + m_2 + am(y_1)$. We must exert caution at this point because a different choice of coordinates on E_α and E_β will lead to a different value for M. In other words, if we wish to give the principal value integral a definition that is independent of the embedding of E_α and E_β in Y_Γ then we must place restrictions on the coordinates used to compute the value of M. It is easy to list some conditions on the coordinates that will insure that M is well-defined.

As above, at any point $p \in E_\alpha(u,F) \cap E_\beta(u,F)$ we begin by selecting local analytic coordinates that are the restriction to $E_\alpha(u,F)$ and $E_\beta(u,F)$ of local analytic coordinates on Y_Γ near p. We also let α be the restriction to $E_\alpha(u,F) \cap E_\beta(u,F)$ of the function defined on a patch by

$$\alpha = \lambda/(\mu_1\mu_2\mu_3^a).$$

Shrinking the patches U_1 and U_2 if necessary, any other system of coordinates $(x_1', y_1'), (x_2', y_2')$ together with a function α' on $U_1 \cap U_2$ must then satisfy (Ai-iv) together with the following conditions on neighborhoods U_1 and U_2 of $E_\alpha(u,F)$ and $E_\beta(u,F)$:
B.i) $x_i/x_i' = \varphi_i$ where φ_i is regular and invertible on $U_1 \cap U_2$.
B.ii) $y_i/y_i' = \psi$ where ψ is regular and invertible on $U_1 \cap U_2$. (By A.i ψ is independent of i.)
B.iii) $|\varphi_i|$ and $|\psi|$ are constant on $U_1 \cap U_2$,
B.iv) $\alpha'/\alpha = \varphi_1\varphi_2\psi^a$ on $U_1 \cap U_2$.
The definition of M is clearly independent of the choice of coordinates satisfying these conditions on sufficiently small patches. We have:

$$m' = m + m(\varphi_1) + m(\varphi_2) + am(\psi)$$

$$m_1' = m_1 - m(\varphi_1)$$

$$m_2' = m_2 - m(\varphi_2)$$

$$am(y_1') = am(y_1) - am(\psi).$$

REMARK 3.1. To specify the principal value integrals we must select a system of coordinates satisfying the conditions listed above. Such a system relates the scale of regions in E_α to the scale on E_β. The form ω_{E_α} (resp. ω_{E_β}) fails to provide a scale because near ω_{E_α} it is scale invariant:

$$|\omega_{E_\alpha}(c.x_1, y_1)| = |\omega_{E_\alpha}(x_1, y_1)| \text{ on } U_1.$$

REMARK 3.2. Notice also that by extending the norm $|\ \ |$ to a field extension K/F, it is not necessary to assume that the original coordinates on Y_Γ near p are defined over F. The reason for this is that coordinates over F satisfying the conditions $A.i$-iv, $B.i$-iv above can always be found and the calculation of M is independent of the coordinates over F satisfying these conditions. In section 6 we will give rational functions (called canonical coordinates) on $E_\alpha(u)$ and $E_\beta(u)$ which give local coordinates on $E_\alpha(u,F)$ and $E_\beta(u,F)$ near every $p \in E_\alpha(u,F) \cap E_\beta(u,F)$. These canonical coordinates will thus provide us with a scale between E_α and E_β.

4. Igusa data for interchanged divisors.
It is possible for two divisors to contain F-rational points without themselves being defined over F. This section develops a formula for the contribution of these F-rational points to the asymptotic expansion. The contribution is expressed as a principal value integral over the intersection of two divisors interchanged by the Galois group of a quadratic field extension K of F. Suppose the

two divisors that are interchanged by the Galois group are E_1 and E_2 and that both have Igusa constants $a(E_1) = a(E_2) = 1, b(E_1) = b(E_2) = b$. Suppose that on a Zariski open set U we have a relation

$$\lambda = \alpha_0 x_1 x_2$$

where x_1 is a regular function such that $x_1 = 0$ defines E_1 and x_2 is a regular function such that $x_2 = 0$ defines E_2 and α_0 is regular on U and invertible on open set of $E_1 \cap E_2$.

For every F-rational point on an open subset of $E_1 \cap E_2$, we construct a cocycle a_σ of $H^1(U(1))$ as follows. ($U(1)$ is defined by the quadratic field extension K/F, where K is the field over which E_1 and E_2 are defined.) The cocycle of $H^1(\mathrm{Gal}(K/F), U(1,K))$ given by $\sigma \rightarrow (\lambda)$ pulls back to a cocycle a'_σ in $H^1(U(1))$. This does not extend to $E_1 \cap E_2$ but

$$a'_\sigma \sigma([x_1])[x_1]^{-1}, [x_1] \in U(1, \bar{F})$$

does extend to an open set of $E_1 \cap E_2$. We take this to be our cocycle a_σ. Note that the cohomology class of a_σ is independent of the choice of local coordinates. Let η_K be the non-trivial character of $H^1(U(1))$. Finally we restrict ourselves to the case that f extends to a locally constant function on a Zariski open subset of $E_1 \cap E_2$.

PROPOSITION 4.1. *The contribution of the F-rational points on E_1 is given by*

$$(1/2)|\lambda|^b \int |dX/X| \int h_2|\nu_2| + (1/2)\eta_K(\lambda)|\lambda|^b \int |dX/X| \int \eta_K(a_\sigma)h_2|\nu_2|$$

where ν_2 is the restriction of $(\omega_Y/\lambda^{b+1})dx_1 dx_2/(x_1 x_2)$ to $E_1 \cap E_2, h_2$ is the restriction of f to $E_1 \cap E_2, X$ varies over norm 1 elements in the field K and the second integral is taken over F-rational points in $E_1 \cap E_2$.

PROOF. A more elegant proof of the result in far greater generality could be given using Mellin transforms. I will settle for a direct proof in this special case. As the size of the mesh goes to zero, the principal value integral on each region also goes to zero. This is clear from formulas appearing in [15,p.475]. So by removing a region with arbitrarily small integral we may work exclusively on patches of the underlying p-adic manifold $U(K)$ which satisfy the conditions:
i) U does not intersect any divisors other than E_1 and E_2.
ii) $\mu_1 = 0$ defines E_1 and $\mu_2 = 0$ defines E_2.
iii) f is locally constant on $U(F)$.
iv) $\omega_Y = \gamma \mu_1^b \mu_2^b d\mu_1 d\mu_2 \ldots .d\mu_n$ with $|\gamma|$ constant on U.
v) $U(K) = \{(\mu_1, \ldots, \mu_n) : |\mu_i| \leq q^{-m_i}\}$ and $m_1 = m_2$.
vi) $U(F) = \{(\mu_1, \ldots, \mu_n) \in U(K) : \sigma(\mu_1) = \mu_2, \sigma(\mu_i) = \mu_i \; i \geq 3\}$

$$(\sigma \in \mathrm{Gal}(K/F), \sigma \neq 1).$$

vii) $\lambda = \alpha\mu_1\mu_2$ with $|\alpha|$ constant on $U(K)$.

We drop the factor γ from the differential form because $|\gamma|$ is constant on $U(K)$. We also ignore the function f because it is locally constant.

If λ/α is not a norm then there are no rational points in the region of integration and the contribution is zero. To compensate for this we insert the function $(1+\eta_K(\lambda/\alpha))/2$ which vanishes precisely when λ/α is not a norm. Every $x \in F^\times$ sufficiently close to the identity is a norm of an element in $K^\times$. It follows that $\eta_K(\alpha(\mu_1, \sigma(\mu_1), \mu_3, \ldots, \mu_n)) = \eta_K(\alpha(0, 0, \mu_3, \ldots, \mu_n))$ for sufficiently small μ_1. Thus we may assume that $\eta_K(1/\alpha)$ is independent of μ_1 and μ_2. We must check that $\eta_K(1/\alpha)$ equals η_K evaluated on the cocycle a_σ. $1/\alpha \in F^\times$ lifts to the cocycle c_σ in $Z^1(U(1))$ given by $\sigma \to 1$, $\sigma|_K = 1, \sigma \to 1/\alpha, \sigma|_K \neq 1$. For λ small but nonzero it has the same class as $\sigma([\mu_1])[\mu_1]^{-1}c_\sigma \quad [\mu_1] \in U(1, K)$. Thus c_σ has the same class for small non-zero λ as the cocycle $\sigma \to 1, \sigma|_K = 1, \sigma \to 1/\lambda, \sigma|_K \neq 1$. It is now clear that $\eta_K(1/\alpha) = \eta_K(a_\sigma)$ on F-rational points.

The region of integration is given by $|\mu_i| \leq q^{-m_i}$ $i = 1, \ldots, n$. $i = 1, 2$ gives $|\lambda/\alpha\mu_2|, |\mu_2| \leq q^{-m_1}$ or

$$(*) \qquad |\lambda/\alpha| q^{m_1} \leq |\mu_2| \leq q^{-m_1}.$$

When $\mu_2\sigma(\mu_2) = \lambda/\alpha$ and λ is sufficiently small the inequalities $(*)$ always hold, so we integrate over all μ_2 with $\mu_2\sigma(\mu_2) = \lambda/\alpha$. When λ/α is a norm select $x \in K^\times$ such that $x\sigma(x) = \lambda/\alpha$ and set $\mu_2 x^{-1} = \mu$. The integral now extends over all norm 1 elements.

The form $\omega_Y/(\lambda^b d\lambda)$ equals by (vii) and (iv)

$$(1/\alpha)^b (d(\lambda/\alpha)/d\lambda)(d\mu_2/\mu_2) \wedge d\mu_3 \ldots d\mu_n =$$

$$(*) \qquad (1/\alpha)^b (d(\lambda/\alpha)/d\lambda)(d\mu/\mu) \wedge d\mu_3 \ldots d\mu_n.$$

For sufficiently small λ, $|d(\lambda/\alpha)/d\lambda| = 1/|\alpha|$. If we integrate out the dependence $d\mu/\mu$ on the norm 1 elements, then for sufficiently small λ the norm of the form $(*)$ equals the norm of the form

$$(1/\alpha)^{b+1} d\mu_3 \ldots d\mu_n = \omega_Y/(\alpha\lambda^b d\mu_1 d\mu_2) = \omega_Y \mu_1\mu_2/(\lambda^{b+1} d\mu_1 d\mu_2).$$

Also $\omega_Y \mu_1\mu_2/(\lambda^{b+1} d\mu_1 d\mu_2)$ restricted to $E_1 \cap E_2$ equals the restriction of $\omega_Y x_1 x_2/(\lambda^{b+1} dx_1 dx_2)$. This proves the lemma.

5. Transition functions.

We have seen that on an open patch $U(\alpha, \beta)$ we can introduce coordinates $w = w(\alpha+\beta)$ and ξ. Fix two lines ℓ_α and ℓ_β of $(B\backslash G)_u$ that intersect at B_+ and select a coordinate patch (B_0, B_∞) with $B_0 = B_+$. Let $E(\ell_\alpha, u)$ and $E(\ell_\beta, u)$ be the components of $E_\alpha(u)$ and $E_\beta(u)$ corresponding to these two lines. This section considers the question of what the coordinate transition functions are when two coordinate patches overlap. First we consider the effect of fixing B_0 and varying B_∞. Suppose we have pairs (B_0, B_∞) and $(B_0, B'_\infty) = (B_0, B_\infty)^n, n \in N_0$. Stars are related by $(B_0^{n_w\nu}) = (B_0^{n'_w\nu'})$ with $n'_w, \nu' \in N'_\infty = N_\infty^n$. Write $n'_w = n''^n_w, \nu' = \nu''^n$ with $n''_w, \nu'' \in N_\infty$. Then $B_0 n_w \nu = B_0 n''_w \nu'' n$. We add double primes to all functions on (B_0, B'_∞).

LEMMA 5.1. *Suppose* $n^{-1} = \exp(yX_\alpha)$ *modulo* N_α. *Then the following statements are true on* $E(\ell_\alpha, u)$.

a) $\xi'' = \xi/(y\xi + 1)$

b) $\delta''w'' = \delta w/(\alpha(X)y\xi\delta w + (1 + y\xi))$

c) $\delta''(\xi'')^{-1}dw''d\xi''/(w''^2\xi'') = \delta(\xi)^{-1}dwd\xi/(w^2\xi)$

Furthermore, the following statements hold on $E(\ell_\alpha, u) \cap E(\ell_\beta, u)$.

d) $\xi''/\xi = 1$

e) $x''(W, \alpha')/x(W, \alpha') = 1$ *for any simple root* α' *and any Weyl chamber* W

f) $w''/w = 1$

(Recall that $\delta(\xi)$ *appearing in (b) and (c) is defined by* $\delta(\xi)\xi = x(\beta)/x(\gamma)$.*)*

PROOF. $n_w\nu = \exp((a_w + \xi)X_{-\alpha})$ on $E(\ell_\alpha, u)$ (a_w is defined in the proof of 1.2) so the 2×2 matrix calculation

$$\begin{aligned}\begin{pmatrix}1 & 0\\ x & 1\end{pmatrix}\begin{pmatrix}1 & y\\ 0 & 1\end{pmatrix} &= \begin{pmatrix}1 & y\\ x & 1+xy\end{pmatrix}\\ &= \begin{pmatrix}1/(xy+1) & y\\ 0 & xy+1\end{pmatrix}\begin{pmatrix}1 & 0\\ x/(xy+1) & 1\end{pmatrix}\end{aligned}$$

shows that $B_0 n_w \nu n^{-1} = B_0 \exp(((a_w+\xi)/(y(a_w+\xi)+1))X_{-\alpha})m_1$ with $m_1 \in N_\alpha$. So $B_0 n_w \nu n^{-1} = B_0 \exp(((a_w + \xi)/(y(a_w + \xi) + 1))X_{-\alpha})$ (for N_α is normal in P_α). Thus $a''_w + \xi'' = (a_w + \xi)/(y(a_w + \xi) + 1)$. In particular for $W = W_+$ we obtain (*a*) $\xi'' = \xi/(y\xi + 1)$.

Let $W = W(\sigma_\alpha)$ so $a_w = z(W_+, \alpha)$. Then $a''_w + \xi'' = (a_w + \xi)/(y(a_w + \xi) + 1)$ becomes

$$(z''(W_+, \alpha)/\xi'' + 1)\xi'' = (z(W_+, \alpha)/\xi + 1)\xi/(y(z(W_+, \alpha)/\xi + 1)\xi + 1).$$

Now by (2.4), $z(W+, \alpha)/\xi = \delta(\xi)w\alpha(X)$ and similarly

$$z''(W_+, \alpha)/\xi'' = \alpha(X)w''\delta''(\xi'').$$

Using (*a*) we obtain

$$(\delta''w''\alpha(X) + 1)\xi'' = (\alpha(X)w\delta + 1)\xi/(y(\alpha(X)w\delta + 1)\xi + 1)$$

$$\begin{aligned}\delta''w''\alpha(X) + 1 &= (1 + y\xi)(\alpha(X)w\delta + 1)/(y\xi\alpha(X)w\delta + (1 + y\xi))\\ \delta''w''\alpha(X) &= \alpha(X)w\delta/(y\xi\alpha(X)w\delta + (1 + y\xi)).\end{aligned}$$

This proves (*b*).

If $w'' = (aw+b)/(cw+d)$ then $dw''/w''^2 = (ad - bc)dw/(aw+b)^2$. So holding ξ, ξ'' constant $\delta''^{-1}dw''/w''^2 = \delta^{-1}(y\xi + 1)dw/w^2$. Also $d\xi''/\xi'' = d\xi/(y\xi + 1)\xi$ and (*c*) follows.

In $E(\ell_\alpha, u) \cap E(\ell_\beta, u)$, $B(W) \in \ell_\alpha \cap \ell_\beta$ for all W. Thus $B(W) = B_0$ for all W. On $E(\ell_\alpha, u), B(W) = B_0^{n_w\nu}, n_w\nu = \exp((a_w + \xi)X_{-\alpha})$. Thus $a_w + \xi = 0$ for all W on $E(\ell_\alpha, u) \cap E(\ell_\beta, u)$. In particular for $W = W_+$ we see that $\xi = 0$. Thus (d) follows from (a).

(e) If $\alpha' \neq \alpha, \beta$ then the result is trivial. In fact for any positive root γ such that $x_\gamma(u) \neq 0$ we have the following string of equalities: $x_\gamma(u) = x_\gamma(u^{\nu^{-1} n_w^{-1}}) = x(W, \gamma) = x_\gamma(u^{\nu'^{-1} n_w'^{-1}}) = x''(W, \gamma)$. (We use $n_w \nu = n'_w \nu' = 1$ on $E(\ell_\alpha, u) \cap E(\ell_\beta, u)$.)

Suppose $\alpha' = \alpha$. $a''_w + \xi'' = (a_w + \xi)/(y(a_w + \xi) + 1), \xi'' = \xi/(y\xi + 1)$. Thus

$$a''_w = (a_w + \xi)/(y(a_w + \xi) + 1) - \xi/(y\xi + 1) =$$

$$a_w(1 + y\xi)^{-1}(1 + y(a_w + \xi))^{-1}.$$

Thus $a''_w/a_w = 1$ on $E(\ell_\alpha, u) \cap E(\ell_\beta, u)$. Selecting $W = W(\sigma_\alpha)$ we obtain $z''(W_+, \alpha)/z(W, \alpha) = 1$ on $E(\ell_\alpha, u) \cap E(\ell_\beta, u)$. The relations

$$\lambda T(W, \alpha) = z''(W_+, \alpha) x''(W_+, \alpha)$$
$$\lambda T(W, \alpha) = z(W, \alpha) x(W, \alpha)$$

now imply that $x''(W, \alpha)/x(W, \alpha) = 1$ on $E(\ell_\alpha, u) \cap E(\ell_\beta, u)$. By interchanging the roles of α and β we obtain the proof for $\alpha' = \beta$.

(f) Set

$$E_{\alpha,\beta,u} = E(\ell_\alpha, u) \ \cap \ E(\ell_\beta, u).$$

By (b) $\delta'' w''/(\delta w) = 1/(\alpha(X)\xi\delta w y + (1 + y\xi)) = 1$ on $E_{\alpha,\beta,u}$. Thus (f) follows if and only if $\delta''/\delta = 1$ on $E_{\alpha,\beta,u}$. On $E_{\alpha,\beta,u}$

$$\delta''/\delta = \delta''\xi''/(\delta\xi) = x''(\beta)x(\gamma)/(x(\beta)x''(\gamma)) = x(\gamma)/x''(\gamma) = 1.$$

The first equality is a result of (d), the second holds by definition, the third equality is a result of (e), and the last equality follows from the string of equalities at the beginning of the proof of (e).

Now we turn to the situation where ℓ_α intersects at least two other lines ℓ_β and $\tilde{\ell}_{\beta'}$ (β and β' not necessarily distinct) with corresponding Borel subgroups B_+ and B_- respectively.

LEMMA 5.2. *Suppose that a line ℓ_α of $(B\backslash G)_u$ intersects at least two other lines of $(B\backslash G)_u$. Let B_- and B_+ be two different Borel subgroups determined by these intersections. Suppose that the set $\{B_+, B_-\}$ is fixed by* $\mathrm{Gal}(\bar{F}/F)$. *Then $B_- \cap B_+$ contains a Cartan subgroup T_0 which is defined over F. Also $B_-^{\sigma_\alpha} = B_+$ where σ_α is the simple reflection in the Weyl group of T_0 corresponding to the simple root α.*

REMARK. We will see in the proof that T_0 depends on the choice of a Levi component M_α in P_α and that the various choices of T_0 are conjugate by $N_\alpha(F)$.

PROOF. The Borel subgroups in ℓ_α fill out a parabolic subgroup P_α which is defined over F because $\mathrm{Gal}(\bar{F}/F)$ fixes ℓ_α. (As always we are working in a perfect field.) Let M_α be a Levi component of P_α which is defined over F. M_α has semi-simple rank one so that the intersection of any two distinct Borel subgroups of M_α is a Cartan subgroup T. Thus $B_+ \cap M_\alpha \cap B_- \cap M_\alpha = B_+ \cap B_- \cap M_\alpha$ is a Cartan subgroup of M_α and hence of G. T_0 is defined over F because M_α is defined over F and B_+, B_- are either fixed or interchanged by the Galois group $\mathrm{Gal}(\bar{F}/F)$ so $B_+ \cap B_-$ is also defined over F.

Since $B_+ \cap M_\alpha$ and $B_- \cap M_\alpha$ are opposite in M_α with intersection T_0, we have

$$(B_+^{\sigma_\alpha} \cap M_\alpha) = (B_+ \cap M_\alpha)^{\sigma_\alpha} = B_- \cap M_\alpha.$$

Two Borel subgroups in P_α are equal if and only if their intersections with M_α are equal. Thus $B_+^{\sigma_\alpha} = B_-$.

By the previous lemma we have a natural patches on $E(\ell_\alpha, u)$ given a pair of Borel subgroups B_+ and B_- lying at the intersections of two lines. If B_+ lies in ℓ_β on $U(\alpha,\beta)$ we let $B_0 = B_+$ and let B_∞ be the Borel subgroup opposite to B_+ through T_0. Similarly if B_- lies in $\tilde{\ell}_{\beta'}$, then on $U(\alpha,\beta')$ we let $B_0' = B_-$ and let B_∞' be the Borel subgroup opposite to B_- through T_0. We relate the two pairs of coordinates on the intersection of the two patches. Let ω_α be an element of the normalizer of T_0 which represents the simple reflection σ_α in the Weyl group. If $(b, (B_0^{n_w}))^\nu = (b', (B_-^{n'_w}))^{\nu'}$ where $b \in B_0 = B_+$, $n_w, \nu \in N_\infty$, $b' = b''^{\omega_\alpha} \in B_- = B_0^{\omega_\alpha}$, $n'_w = n''^{\omega_\alpha}_w \in N_\infty^{\omega_\alpha}$, $\nu' = \nu''^{\omega_\alpha} \in N_\infty^{\omega_\alpha}$, then clearly $(b, B_0^{n_w})^\nu = (b'', B_0^{n''_w})^{\nu''\omega_\alpha}$, with $b, b'' \in N_\alpha$ and $n_w, n''_w, \nu, \nu'' \in N_\infty$. Let (w,ξ) and (w'', ξ'') be coordinates on these two patches.

LEMMA 5.3.
a) $\xi = 1/\zeta\xi''$ *for some* $\zeta \in \bar{F}^\times$ *depending on* ω_α.
b) $\delta w = -\delta'' w''/(\delta''\alpha(X)w'' + 1)$.

PROOF. Write

$$n_w = \exp(a_w X_{-\alpha})^\alpha n_w, \quad n''_w = \exp(a''_w X_{-\alpha})^\alpha n''_w$$
$$\nu = \exp(\xi X_{-\alpha})^\alpha \nu, \quad \nu'' = \exp(\xi'' X_{-\alpha})^\alpha \nu''$$

with ${}^\alpha n_w$, ${}^\alpha n''_w$, ${}^\alpha \nu$, ${}^\alpha \nu'' \in N_{-\alpha}$. Then

$$n''_w \nu'' \omega_\alpha = \exp((a''_w + \xi'')X_{-\alpha})\omega_\alpha \nu_1$$

where $\nu_1 \in N_{-\alpha}$. By the 2×2 matrix calculation

EQUATION 5.4.

$$\begin{pmatrix} 1 & 0 \\ x & 1 \end{pmatrix} \begin{pmatrix} 0 & a \\ b & 0 \end{pmatrix} = \begin{pmatrix} 0 & a \\ b & ax \end{pmatrix} = \begin{pmatrix} -b/x & a \\ 0 & ax \end{pmatrix} \begin{pmatrix} 1 & 0 \\ b/(ax) & 1 \end{pmatrix}$$

we see that

$$\begin{aligned} B_0 n''_w \nu'' \omega_\alpha &= B_0 \exp((1/\zeta(a''_w + \xi''))X_{-\alpha})\nu_1 \\ &= B_0 n_w \nu = B_0 \exp((a_w + \xi)X_{-\alpha})\nu_2, \end{aligned}$$

$\nu_2 \in N_{-\alpha}, \zeta = a/b$. Thus $1/\zeta(a''_w + \xi'') = (a_w + \xi)$. In particular, taking $W = W_+$ we obtain $1/\zeta\xi'' = \xi$. This proves (a).

(b) $\xi/(a_w + \xi) = (a''_w + \xi'')/\xi''$ or $(a_w/\xi) + 1 = 1/((a''_w/\xi'') + 1)$.

Let $W = W(\sigma_\alpha)$ so that $a_w = z(W_+, \alpha)$ and $a''_w = z''(W_+, \alpha)$. Now by (V.2.4), $z(W_+, \alpha)/\xi = \alpha(X)\delta w$. Similarly $z''(W_+, \alpha)/\xi'' = \alpha(X)w''\delta''$. Thus

$$\alpha(X)w\delta + 1 = (\alpha(X)w''\delta'' + 1)^{-1} \text{ or}$$

$$\delta w = -\delta'' w''/(\alpha(X)w''\delta'' + 1).$$

6. Coordinate Relations.

For any two adjacent roots α_1 and α_2 we define a constant $e = e(\alpha_1, \alpha_2)$ by the condition

$$\exp(X_{-\alpha_1})\exp(X_{\alpha_1+\alpha_2})\exp(-X_{-\alpha_1}) = \exp(eX_{\alpha_2}) \text{ modulo } N_{\alpha_2}.$$

Recall from chapter II that $x(W, \alpha) = x_\alpha(b^{n_w^{-1}})$, and that $w(\gamma)$ depends on an ordering on the positive roots.

LEMMA 6.1. *Suppose that ℓ_α intersects a line ℓ_β at B_+ in $(B\backslash G)_u$. Let the Borel subgroup B_0 defining a coordinate patch (B_∞, B_0) be given by $B_0 = B_+$. If α is adjacent to a long root then we assume that ℓ_β corresponds to a long root. If ℓ_α intersects a second line we also require that T_0 be chosen so that $B_0^{\sigma_\alpha}$ lies in the intersection of ℓ_α and a second line $\tilde{\ell}_{\beta'}$. (Cf. 5.2). If β is longer than α we require that B_0 and T_0 are chosen so that $\beta' = \beta$. Finally we exclude the group G_2 when β is longer than α. On this coordinate patch the following statements hold in the coordinate ring of $E(\ell_\alpha, u)$. They hold independent of the implicit ordering on the roots.*

a) If α' is longer than α and $(\alpha', \alpha) \neq 0$, then $\alpha' = \beta$ and $x(2\alpha + \beta) = 0$.

b) $x(W, \alpha + \beta) = x(\alpha + \beta) = x_{\alpha+\beta}(u) \;\; \forall\, W$.

c) $\delta(\xi) = e(\alpha, \beta)$.

d) $z(W_+, \alpha)/\xi = \alpha(X)we(\alpha, \beta)$.

e) If $w(\gamma) \neq 0$ and γ is not simple then $\gamma = \alpha + \alpha'$ where $\alpha' \neq \beta'$ if $\beta' \neq \beta$.

f) For any two simple roots α' and α'' (not necessarily distinct),

$$x(W(\sigma_{\alpha''}), \alpha')/x(\alpha') = 1 + e(\alpha'', \alpha')\alpha''(X)w(\alpha' + \alpha'')$$

(Set $w(\alpha' + \alpha'') = 0$ if $\alpha' + \alpha''$ is not a root.)

g) w and ξ are independent of the order selected on the roots. If ℓ_α intersects more than one line, w and ξ depend only on the choice of lines ℓ_β and $\tilde{\ell}_{\beta'}$ and not on the choice of T_0 satisfying the hypotheses of the lemma.

REMARK 1. By $(g)w$ and ξ depend only on the choice of lines ℓ_β and $\tilde{\ell}_{\beta'}$ and are called *canonical* coordinates.

REMARK 2. It should be pointed out that despite the large number of hypotheses, given any line ℓ_α (except the short line in G_2) B_0 and T_0 can be chosen to satisfy the hypotheses.

PROOF. First I show that the equations in $(a), (b)$ do not depend on the ordering on the roots. By rearranging the order of the product we have

$$n = \prod \exp(x(\gamma)X_\gamma) \text{ (fixed order) and}$$
$$n = \prod \exp(y(\gamma)X_\gamma) \text{ (different order)}$$

By $(II.3.1)$, we have identities of the form:

$$x(\gamma) + \sum_{n \geq 2} d_{\beta_1 \ldots \beta_n} x(\beta_1) \ldots x(\beta_n) = y(\gamma)$$

with $\beta_1 + \ldots + \beta_n = \gamma$. The values of the constants $d_{\beta_1 \ldots \beta_n}$ do not concern us here. For simple roots β we obtain $x(\beta) = y(\beta)$. For $\gamma = \alpha + \beta$ we obtain $x(\alpha + \beta) + d_{\alpha\beta} x(\alpha)x(\beta) = y(\alpha + \beta)$. On E_α $x(\alpha) = 0$ so $x(\alpha + \beta) = y(\alpha + \beta)$ – independent of the order. Similarly for $\gamma = 2\alpha + \beta$ we obtain $x(2\alpha + \beta) + d_{\alpha\alpha\beta} x(\alpha)^2 x(\beta) + d_{\alpha\gamma} x(\alpha)x(\gamma) = y(2\alpha + \beta)$. Again on E_α $x(2\alpha + \beta) = y(2\alpha + \beta)$ independent of the order.

(a) The first conclusion is true by hypothesis. Also by hypothesis we are at a node of type B_2. $u \in B_0^{\sigma_\alpha} = B_-$ and B_- also lies in a line of type β. This is so if and only if the βth coefficient of u^{σ_α} is zero, that is the $\sigma_\alpha(\beta)th$ coefficient of u is zero. But if β is longer than α in a node of type B_2 then $\sigma_\alpha(\beta) = 2\alpha + \beta$. So $x(2\alpha + \beta) = 0$.

(b) Write $u = \prod \exp(x_{\alpha'}(u)X_{\alpha'})$. $x_\alpha(u) = x_\beta(u) = 0$ by the choice of B_0. Since u is subregular $x_{\alpha+\beta}(u) \neq 0$ $(V.1.1)$. Now $(u, (B(W)) = (u, B_0^{n_w \nu}) = (u^{\nu^{-1}}, B_0^{n_w})^\nu$. So $b = u^{\nu^{-1}}$ and $b^{n_w^{-1}} = u^{\nu^{-1} n_w^{-1}}$.

$$n_w \nu = \exp((a_w + \xi)X_{-\alpha})$$

on $E_\alpha(u)$ (1.2). So $b^{n_w^{-1}} =$

$$(*) \qquad \exp((\xi + a_w)X_{-\alpha})\{\prod \exp(x_{\alpha'}(u)X'_\alpha)\} \exp(-(\xi + a_w)X_{-\alpha}).$$

We are interested in the $\alpha + \beta th$ coefficient of this product. Suppose first that $|\alpha| \geq |\beta|$. With respect to the Weyl chamber $W(\sigma_\alpha), \alpha + \beta$ and $-\alpha$ are positive simple roots. Therefore $\exp(-(\xi + a_w)X_{-\alpha})$ can be passed to the left in $(*)$ without affecting the $\alpha + \beta th$ coefficient in $\prod \exp(x_{\alpha'}(u)X'_\alpha)$. Now suppose that $|\alpha| < |\beta|$. We work inside B_2. The simple roots with respect to the Weyl chamber $W(\sigma_\alpha)$ are $2\alpha + \beta$ and $-\alpha$. Again $\exp(-(\xi + a_w)X_{-\alpha})$ can be passed to the left in $(*)$ without affecting the $\alpha + \beta th$ coefficient in $\prod \exp(x(\alpha')X_{\alpha'})$. This is because $\alpha + \beta = (2\alpha + \beta) + (-\alpha)$ (as a sum of simple roots with respect to $W(\sigma_\alpha)$) and $x(2\alpha + \beta) = 0$. This proves (b).

(c) I claim first of all that if $\eta = m(\alpha)\alpha + m(\beta)\beta$, $m(\alpha) + m(\beta) \geq 3$ then

$$\exp(\xi X_{-\alpha}) \exp(x(\eta)X_\eta) \exp(-\xi X_{-\alpha}) \in N_\beta.$$

The only interesting case occurs when $\eta = \sigma_\alpha(\beta)$. By (a) we may assume that β is no longer than α. But when β is no longer than $\alpha, \sigma_\alpha(\beta) = \alpha+\beta$ contradicting the hypothesis that $m(\alpha)+m(\beta) \geq 3$.

Select a point $p = (u,(B(W))) \in E_\alpha(u)$ such that $B(W_+) = B_0$. Then $p = (u,(B_0^{n_w}))$. As above $x_\alpha(u) = x_\beta(u) = 0$. Now let $(u,(B(W)))$ be any other point in $E_\alpha(u)$. Then $(u,(B(W)) = (u^{\nu^{-1}},(B_0^{n_w}))^\nu$ and (1.2) implies that $\nu = \exp(\xi X_{-\alpha})$ for some ξ. The coefficient $x(\beta)$ is given by the equation

$$\exp(\xi X_{-\alpha})u\exp(-\xi X_{-\alpha}) = \exp(x(\beta)X_\beta) \text{ modulo } N_\beta.$$

By the previous paragraph

$$\exp(\xi X_{-\alpha})u\exp(-\xi X_{-\alpha}) = \exp(\xi X_{-\alpha})\exp(x_0 X_{\alpha+\beta})\exp(-\xi X_{-\alpha}) = \exp(e(\alpha,\beta)x_0\xi X_\beta) \text{ modulo } N_\beta$$

where $x_0 = x_{\alpha+\beta}(u) = x(\alpha+\beta)$ by (b). Thus $x(\beta) = e(\alpha,\beta)x(\alpha+\beta)\xi$ follows from (b). By definition $x(\beta) = \delta(\xi)x(\alpha+\beta)\xi$. We have seen that $x(\beta)$ and $x(\alpha+\beta)$ are independent of the order. ξ is independent of the ordering on the roots because $B(W_+) = B_0^{\exp(\xi X_{-\alpha})}$ on $E(\ell_\alpha,u)$ independent of the ordering.

(d) This follows immediately from (c) and (2.4).

(e)

$$\lambda w(\gamma) = x(\gamma)\prod z(\alpha')^{m(\alpha')}$$
$$= x(\gamma)\prod(\lambda/x(\alpha'))^{m(\alpha')}z(\alpha)^{m(\alpha)}.$$

Set $m = \sum m(\alpha)$ if $\gamma = \sum m(\alpha)\alpha$. Then

$$w(\gamma) = \lambda^{m-m(\alpha)-1}x(\gamma)z(\alpha)^{m(\alpha)}\prod(1/x(\alpha'))^{m(\alpha')}.$$

Now $z(\alpha)$ and $x(\alpha'), \alpha' \neq \alpha$ are not identically zero on $E_\alpha(u)$. So $w(\gamma) = 0$ if and only if $\lambda^{m-m(\alpha)-1}x(\gamma) = 0$ on $E_\alpha(u)$. If $m-m(\alpha) > 1$ then $\lambda^{m-m(\alpha)-1}x(\gamma) = 0$. So it is enough to check the case $m - m(\alpha) = 1$ (i.e. $\gamma = m(\alpha)\alpha + \alpha'$).

If $m(\alpha) > 1$ then α' is longer than α and it follows from (a) that $x(\gamma) = 0$. So we limit ourselves to the case that $m(\alpha) = 1, \alpha'$ is no longer than α, and $\alpha' = \beta', \beta' \neq \beta$. As in the proof of (a) $u^{\sigma_\alpha} \in B_+^{\sigma_\alpha} = B_-$ on which the $\alpha' th (= \beta' th)$ coefficient is zero. But the $\beta' th$ coefficient of u^{σ_α} is zero if and only if the $\sigma_\alpha(\beta')th$ coefficient of u is zero. $\sigma_\alpha(\beta') = \alpha+\beta'$ so $x(\alpha+\beta') = 0$.

(f) We begin with a few simple cases. Write $n_{w''} = \exp(zX_{-\alpha''})$ with $z = z(W_+,\alpha'')$ and $W'' = W(\sigma_{\alpha''})$.

Case 1. $\alpha' = \alpha''$. We appeal to the 2×2 matrix calculation

$$\begin{pmatrix} 1 & 0 \\ z & 1 \end{pmatrix}\begin{pmatrix} t_1 & t_1 x \\ 0 & t_2 \end{pmatrix}\begin{pmatrix} 1 & 0 \\ -z & 1 \end{pmatrix} = \begin{pmatrix} t_1 & t_1 x \\ zt_1 & zt_1x+t_2 \end{pmatrix}\begin{pmatrix} 1 & 0 \\ -z & 1 \end{pmatrix} =$$
$$\begin{pmatrix} t_1 - zt_1x & t_1x \\ z(t_1-t_2-zt_1x) & zt_1x+t_2 \end{pmatrix}.$$

Now $t_1 - t_2 - zt_1x(\alpha) = 0$ by $(II.6.1)$.

$$= \begin{pmatrix} t_2 & t_1x \\ 0 & t_1 \end{pmatrix} = \begin{pmatrix} t_2 & t_2x' \\ 0 & t_1 \end{pmatrix}$$

where $x = x(\alpha)$ and $x' = x(W(\sigma_{\alpha'}),\alpha')$. So that $x(W'',\alpha')/x(\alpha') = t_1/t_2$ and upon restriction to $E_\alpha(u)$ $x(W'',\alpha')/x(\alpha') = 1$.

CASE 2. $(\alpha', \alpha'') = 0$. $x(\alpha') = x_{\alpha'}(b)$, and $x(W, \alpha') = x_{\alpha'}(b^{n''_w})$. It is clear that on Y^0 $x(W'', \alpha')/x(\alpha') = 1$ so that the same holds true on $E_\alpha(u)$.

CASE 3. $\alpha' \neq \alpha, \alpha'' \neq \alpha$. The hypothesis that $\alpha'' \neq \alpha$ implies that $z = 0$ on $E_\alpha(u)$. The hypothesis that $\alpha' \neq \alpha$ implies that $x(\alpha')$ is not identically zero on $E_\alpha(u)$. It follows easily that $x(\alpha') = x(W'', \alpha')$ on $E_\alpha(u)$.

Before proceeding to the final cases we prove a preliminary result. Write

$$(*) \qquad \exp(zX_{-\alpha''})\exp(x(\eta)X_\eta)\exp(-zX_{-\alpha''}) = \exp(cX_{\alpha'}) \text{ modulo } N_{\alpha'}$$

where $\eta = m(\alpha')\alpha' + m(\alpha'')\alpha''$ with $m(\alpha') + m(\alpha'') \geq 3$. c here is a function of z and $x(\eta)$. I claim that the rational function $c/x(\alpha')$ is zero on E_α. c is identically zero unless $\eta = m(\alpha'')\alpha'' + \alpha'$. The conditions $m(\alpha') + m(\alpha'') \geq 3$ and $\eta = m(\alpha'')\alpha'' + \alpha'$ are incompatible unless α'' is shorter than α'. So we assume that α'' is shorter than α'. If $\alpha' \neq \alpha, \alpha'' = \alpha$ then $x(\eta) = 0$ by (a) (using the fact that α'' is shorter than α'). If $\alpha' = \alpha$, and $\alpha'' \neq \alpha$ then conjugate the relation $(*)$ by $t \in T_0$ such that $\alpha(t) = 1/x(\alpha), \alpha''(t) = x(\alpha)$. The relation $(*)$ modulo N_α becomes

$$\exp((z/x(\alpha))X_{-\alpha''})\exp(x(\alpha)^j x(\eta)X_\eta)\exp(-(z/x(\alpha))X_{-\alpha''}) =$$

$$\exp((c/x(\alpha))X_\alpha) \text{ modulo } N_\alpha.$$

Since α'' is shorter than $\alpha', j = m(\alpha'') - m(\alpha') > 0$ so that $x(\alpha)^j x(\eta) = 0$ on E_α. Thus the point will follow if $z/x(\alpha)$ is regular on E_α. $(1 - \alpha''^{-1}) = z(W_+, \alpha'')x(\alpha'')$ and $(1 - \alpha^{-1}) = z(W_+, \alpha)x(\alpha)$ so

$$z/x(\alpha) = (1 - \alpha''^{-1})z(W_+, \alpha)/(x(\alpha'')(1 - \alpha^{-1}))$$

which is regular on an open set of E_α.

Now we move to the proof of (f). $x(W(\sigma_{\alpha''}), \alpha')$ is given by

$$t_0 \exp(zX_{-\alpha''})b\exp(-zX_{-\alpha''}) = \exp(x(W(\sigma_{\alpha''}), \alpha')X_{\alpha'}) \text{ mod } N_{\alpha'},\ t_0 \in T_0.$$

If b is expressed as a product $t \cdot \prod \exp(x(\gamma)X_\gamma)$ then $x(W(\sigma_{\alpha''}), \alpha')/x(\alpha')$ on $E(\ell_\alpha, u)$ becomes a sum $\sum(c_\gamma/x(\alpha'))$ where c_γ is defined by

$$\exp(zX_{-\alpha''})\exp(x(\gamma)X_\gamma)\exp(-zX-_{\alpha''}) = \exp(c_\gamma X_{\alpha'}) \text{ mod } N_{\alpha'}.$$

By the previous paragraph $c_\gamma/x(\alpha') = 0$ except possibly when $\gamma = \alpha'$ or $\alpha' + \alpha''$. (It is clear that $c_\gamma/x(\alpha')$ is zero if $\gamma = \alpha''$.) Now

$$\exp(zX_{-\alpha''})\exp(x(\alpha')X_{\alpha'})\exp(-zX_{-\alpha''}) = \exp(x(\alpha')X_{\alpha'}) \text{ modulo } N_{\alpha'}$$

and

$$\exp(zX_{-\alpha''})\exp(x(\alpha' + \alpha'')X_{\alpha'+\alpha''})\exp(-zX_{-\alpha''}) = \exp(cX_{\alpha'}) \text{ modulo } N_{\alpha'}$$

where

$$c/x(\alpha') = e(\alpha'', \alpha')z(W_+, \alpha'')x(\alpha' + \alpha'')/x(\alpha')$$

by the definition of $e(\alpha'', \alpha')$. Using

$$z(W_+, \alpha'') = T(W_+, \alpha'')\lambda/x(\alpha'') \text{ and } \lambda = x(\alpha')x(\alpha'')w(\alpha' + \alpha'')/x(\alpha' + \alpha'')$$

we obtain

$$c/x(\alpha') = e(\alpha'', \alpha')\alpha''(X)w(\alpha' + \alpha'').$$

This proves (f).

(g) The independence of ξ was observed in the proof of (c). (f) gives

$$x(W(\sigma_\alpha), \beta) = (1 + e(\alpha, \beta)\alpha(X)w)x(\beta).$$

The independence of w of the order now follows from the independence of $x(\beta)$ and $x(W(\sigma_\alpha), \beta)$ of the ordering. The various choices of T_0 are conjugate by any element $n \in N_\alpha$. (5.1) can be applied with $y = 0$, (5.1.a) gives $\xi'' = \xi$, (5.1.b) gives $\delta'' w'' = w\delta$, but from (c) we see that $\delta = \delta'' = e(\alpha, \beta)$, so that $w'' = w$.

The next lemma supplements (6.1) in the case that ℓ_α intersects only one line.

LEMMA 6.2. *Suppose that ℓ_α intersects exactly on line ℓ_β. Select Borel subgroups so that $B_0^{\sigma_\alpha} = B_+$ where B_+ is the Borel subgroup at the intersection of ℓ_α and ℓ_β and σ_α is a simple reflection in the Weyl group of $T_0 = B_0 \cap B_\infty$. Then the following relations hold for functions in the coordinate ring of $E(\ell_\alpha, u)(B_\infty, B_0)$.*
a) $x(\alpha + \beta) = 0$
b) $x(W, \alpha')/x(\alpha') = 1$ for all simple roots α' and all W.
c) $w(\gamma) = 0 : \gamma$ not simple.
These relations hold independent of the implicit ordering on the roots.

PROOF. It follows from the fact that α intersects only one line that $|\alpha| \geq |\beta|$. Thus $\sigma_\alpha(\beta) = \alpha + \beta$. These relations are independent of the ordering on the roots for the same reasons they were in the previous lemma.

(a) $B_0^{\sigma_\alpha} = B_+$ implies that $u^{\sigma_\alpha} \in B_0$ and that the αth and βth coefficients of u^{σ_α} are zero, so the αth and $\sigma_\alpha(\beta)th = \alpha + \beta th$ coefficients of u are zero.

We will return to (b) after proving (c).

$$\begin{aligned}\lambda w(\gamma) &= x(\gamma)\prod z(\alpha')^{m(\alpha')}\\ &= x(\gamma)\prod(\lambda/x(\alpha'))^{m(\alpha')}z(\alpha)^{m(\alpha)}.\end{aligned}$$

Set $m = \sum m(\alpha)$ if $\gamma = \sum m(\alpha)\alpha$. Then

$$w(\gamma) = \lambda^{m-m(\alpha)-1}x(\gamma)z(\alpha)^{m(\alpha)}\prod(1/x(\alpha'))^{m(\alpha')}.$$

$w(\gamma) = 0$ on $E_\alpha(u)$ if $m - m(\alpha) - 1 > 0$. Assume $m = m(\alpha)+1$. $\gamma = m(\alpha)\alpha + \alpha'$. Since ℓ_α intersects only one line $\alpha' = \beta$. β is not longer than α so we must have $\gamma = \alpha + \beta$. Now

$$w(\alpha + \beta) = x(\alpha + \beta)z(\alpha)/x(\beta) = 0$$

because $x(\alpha+\beta)=0$.

(*b*)

$$(*) \qquad \begin{aligned} \lambda T(W,\alpha') &= z(W,\alpha')x(W,\alpha') \\ \lambda T(W',\alpha') &= z(W',\alpha')x(W',\alpha') \end{aligned}$$

so (*b*) will follow if we show $z_1(W,\alpha')/z_1(W',\alpha') = T(W,\alpha')/T(W',\alpha')$ on $E_\alpha(u)$. On the regular divisor E_0, $x(\alpha') = x(W,\alpha')$ independent of W so that $(*)$ implies that
$z_1(W,\alpha')/z_1(W',\alpha') = T(W,\alpha')/T(W',\alpha')$ on E_0. On E_0 we also have

$$w(\gamma) = \lambda^{m-1}x(\gamma)\prod(1/x(\alpha'))^{m(\alpha')}$$

so that $w(\gamma)=0$ for γ not simple. Proposition $II.4.1$ shows that

$$z_1(W,\alpha')/z_1(W',\alpha')$$

lies in the ring generated by λ and $\{w(\gamma)\}$. Since $w(\gamma)$ (γ not simple) and λ are zero on both E_0 and $E_\alpha(u)$ it follows that $z_1(W,\alpha')/z_1(W',\alpha')$ has the same value on $E_\alpha(u)$ as on E_0. This completes the proof.

VI. RATIONALITY AND CHARACTERS

1. Rationality.

This section investigates the rationality structure of the variety $E(\ell_\alpha, u)$. The variables (w, ξ) are not in general defined over F. The rationality structure is determined by the action of the group $\mathrm{Gal}(\bar{F}/F)$ on the coordinates. First we determine the action of $\mathrm{Gal}(\bar{F}/F)$ on the divisors.

LEMMA 1.1. *The Galois group acts on the divisors by $\sigma(E_\alpha) = E_{\sigma*\alpha}$ where the action σ_* on the simple roots is that governed by the quasi-split form of G.*

PROOF. Fix a subregular unipotent element $u \in G(F)$. Let $E_0(u) = \pi^{-1}(u) \cap E_0$. On E_0, $z(\alpha) = 0 \ \forall \ \alpha$. So $n_w = 1 \ \forall \ W$ and

$$B(W) = B(W') \ \forall \ W, W' \text{ if } (u, (B(W))) \in E_0(u).$$

It is easy to see that $(u, (B(W))) \in E_0(u) \cap E_\alpha$ if and only if $B(W) \in \ell_\alpha$ for all W. This gives an isomorphism over F of $E_0(u)$ with $(B\backslash G)_u$. If $B\backslash P_\alpha g$ is a line of type α in $(B\backslash G)_u$ then $\sigma(B\backslash P_\alpha g) = (\sigma(B)\backslash P'_{\sigma*\alpha}\sigma(g))$ where $P'_{\sigma*\alpha}$ is the parabolic subgroup of type $\sigma_*\alpha$ containing $\sigma(B)$. So $\sigma(B\backslash P_\alpha g)$ is a line of type $\sigma_*\alpha$ in $(B\backslash G)_u$. Since the map $E_0(u) \rightarrow (B\backslash G)_u$ is defined over F, the divisor $\sigma(E_\alpha)$ must be associated with the root $\sigma_*\alpha$.

We make three important remarks.

REMARK 1.2. It follows by glancing at the possible graph automorphisms that if a subregular divisor has an F-rational point which projects to a subregular element in $G(F)$ then the divisor itself is defined over F, with the exception of the group ${}^2A_{2n}$ and the divisors E_{α_n} and $E_{\alpha_{n+1}}$. They are not defined over F but their intersection is.

REMARK 1.3. The action of the Galois group $\mathrm{Gal}(\bar{F}/F)$ on the components $\{E(\ell_\alpha, u)\}$ of $E_\alpha(u)$ must also be compatible with the action of $\mathrm{Gal}(\bar{F}/F)$ on the lines of type α in the Dynkin curve $(B\backslash G)_u$. It is important to note that this action will depend on the choice of subregular element $u \in G(F)$ (but certainly not on the Cartan subgroup T). For example, consider the group of type B_n. The Dynkin curve has the form

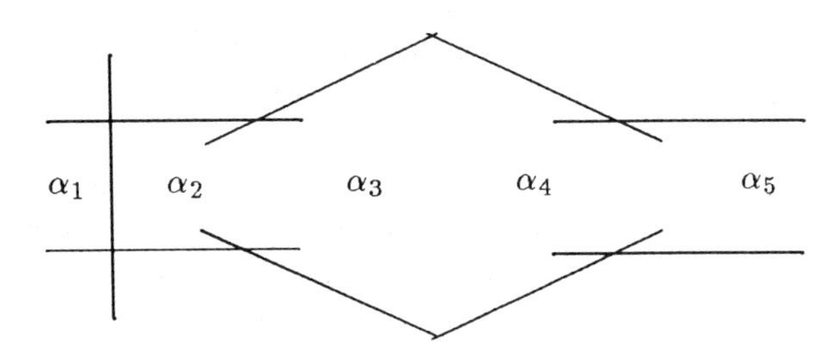

where $|\alpha_1| < |\alpha_i|$ for $i > 1$. It can be shown that for a quasi-split group of type B_n the adjoint conjugacy classes of subregular unipotent elements in $G(F)$ are in $1-1$ correspondence with $F^\times/F^{\times 2}$ where $x \in F^\times/F^{\times 2}$ corresponds to a class O where the field of definition of the lines of $(B\backslash G)_u$ is $F(\sqrt{x})$. There is no *a priori* reason to expect the germs corresponding to unipotent classes with different actions on the Dynkin curve to be related.

REMARK 1.4. If $u \in G(F)$ then $(B\backslash G)_u$ is defined over F. If $(B\backslash G)_u$ contains a fixed point under $\mathrm{Gal}(\bar{F}/F)$ then we conclude it is quasi-split. By examining the various Dynkin curves in diagram $(IV.1)$, we see that the Dynkin curve has a fixed point (marked by "o" in the diagram) at the intersection of two lines under the action of $\mathrm{Gal}(\bar{F}/F)$ except possibly for the Dynkin diagrams and Galois actions: $G_2, {}^3D_4, {}^6D_4, B_n, {}^2A_{2n+1}$. Furthermore, by Kneser's classification [13] groups of type $G_2, {}^3D_4$, and ${}^6D_4$ are always quasi-split. Thus if G is not quasi-split and $G(F)$ contains a subregular unipotent element then the Dynkin curve has the following form.

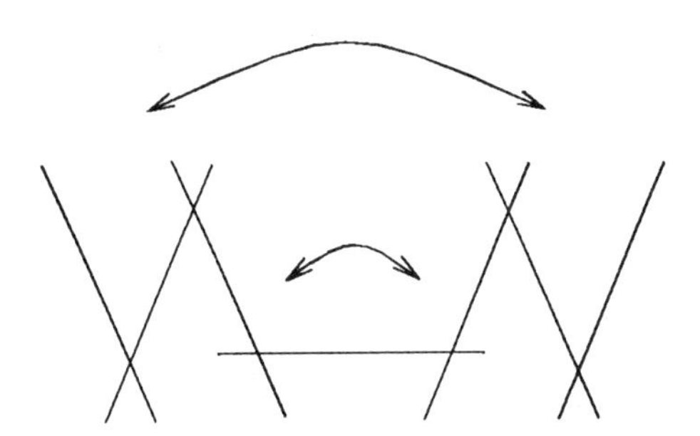

The Galois group exchanges lines as indicated by the arrows; and the line fixed by the Galois group has no F-rational points.

Now we turn to the question of the action of the Galois group on the coordinates (w, ξ). We begin with a split form G_{sp} of G with split Cartan subgroup and Borel subgroup $T_{sp} \subseteq B_{sp} \subseteq G_{sp}$. Let σ_{sp} denote the action of the Galois group on G_{sp}. Fix root vectors such that $\sigma_{sp}(X_\gamma) = X_\gamma$.

Suppose that ℓ_α is a line of $(B\backslash G)_u$ defined over F. Then it is associated with a parabolic subgroup P_α over F. If ℓ_α intersects only one line then the action is given in (3.2). Excluding these cases and the exceptional cases ${}^3D_4, {}^6D_4, G_2$ we see that the hypotheses of $(V.5.2)$ are satisfied. Thus there exists a Cartan subgroup T_0 over F in P_α such that the two Borel subgroups B_+ and $B_- = B_+^{\sigma_\alpha}$ in P_α containing T_0 lie at the intersection of two lines in $(B\backslash G)_u$. (We choose our lines as in (V.6.1).) We set $B_+ = B_0$ and let B_∞ be the Borel subgroup opposite to B_0 through T_0.

We fix an isomorphism over $\bar{F}$ of G with G_{sp} which carries P_α to $P_{\alpha sp} \supseteq B_{sp}$ and T_0 to T_{sp}. B_+ is carried to a Borel subgroup in $P_{\alpha sp}$ containing T_{sp} so we may also assume that the isomorphism carries B_+ to B_{sp}. We identify G and G_{sp} through this isomorphism and write $\sigma(g) = \sigma_{sp}(\mathbf{ad} w_\sigma a_\sigma(g))$ for $g \in G$ and some $w_\sigma \in N_{P_\alpha}(T_0)_{adj}$, and a_σ an automorphism of $(G_{sp}, B_{sp}, T_{sp}, \{X_\gamma\})$. w_σ is the non-trivial element in the Weyl group $\Omega_{P_\alpha}(T_0)$ if and only if σ interchanges the Borel subgroups B_+ and B_-. By [15] or directly from definitions in chapter I we have for a point p

$$(b, B_0^{n_w})^\nu(\sigma p) = (\sigma(b), \sigma(B_0)^{\sigma(n_{w'})})^{\sigma(\nu)}(p),$$

$W' = \sigma^{-1} W$ or

EQUATION 1.5.

$$(*) \qquad \sigma_{sp}^{-1}(b, B_0^{n_w})^\nu(\sigma p) = \mathbf{ad}\ w_\sigma[(a_\sigma(b), B_0^{a_\sigma(n_{w'})})^{a_\sigma(\nu)}].$$

We analyze two cases separately. First suppose that all the roots are the same length. Then there is only one line in $(B\backslash G)_u$ corresponding to each simple root. B_+ lies in ℓ_α and ℓ_β. B_- lies in ℓ_α and $\ell_{\beta'}$ and $\beta' \neq \beta$. But $\sigma(\beta) = \beta'$ if a_σ is non-trivial, so that B_+ and B_- are interchanged if and only if a_σ is non-trivial. We observed above that w_σ gives the non-trivial element in $N_{P_\alpha}(T_0)_{adj}$ if and only if B_+ and B_- are interchanged. We conclude that w_σ is trivial in Ω_{P_α} if and only if a_σ is trivial. Let K be the field of definition of B_+. Over the extension K, $a_\sigma = 1$ so that G splits over K. Thus we may actually arrange that our identification of G with G_{sp} is defined over K (and is independent of T). In particular we may assume that $w_{\sigma\tau} = w_\tau$ and $w_\sigma = 1$ if $\sigma|_K = 1$. Thus w_σ depends only on the image of Gal$(\bar{F}/F)$ in $\tilde{\Omega}$, the extended Weyl group. We see that w_σ takes on only two values 1 and w_0 as σ ranges over elements of Gal$(\bar{F}/F)$. The right hand side of (1.5) depends on T only through $W' = \sigma^{-1} W$. It depends on Gal$(\bar{F}/F)$ only through its image in the extended Weyl group $\tilde{\Omega}$ (using the action of Gal$(\bar{F}/F)$ on chambers W).

Next we consider the case that there are roots of different lengths. Then G is quasi-split if and only if it is split. So $a_\sigma = 1$ for all σ. Again we may identify G with G_{sp} over the field of definition K of B_+. Once again $w_{\sigma\tau} = w_\tau$ and $w_\sigma = 1$ if $\sigma|_K = 1$. We define the group $\tilde{\Omega}$ to be the direct product of a cyclic group of order two and the Weyl group: $\tilde{\Omega} = \Omega \times \mathbb{Z}/2$. Define a homomorphism φ : Gal$(\bar{F}/F) \to \tilde{\Omega}$ by identifying the cyclic group of order two with Gal(K/F) (if $K \neq F$) and sending σ to $\omega \in \Omega$ with $\sigma^{-1} W = \omega^{-1} W \ \forall\ W$. (If $K = F$ we take the image of φ to lie in $\Omega \subseteq \tilde{\Omega}$.) We see as before by (1.5) that the action of Gal$(\bar{F}/F)$ depends only on of $\tilde{\Omega}$.

If the roots are the same length we let σ_0 denote an outer automorphism in $\tilde{\Omega}$ fixing B_{sp}. If they are not the same length we let $\sigma_0 = (1, \epsilon) \in \Omega \times \mathbb{Z}/2 = \tilde{\Omega}, \epsilon \neq 1$. σ_0 and the simple reflections generate $\tilde{\Omega}$ provided $|\tilde{\Omega} : \Omega| \leq 2$. We have now proved the first two statements of

THEOREM 1.6. *Exclude as above the cases of ℓ_α intersecting only one line or three lines with no rational points.*

a) *The isomorphism of G with G_{sp} described above can be chosen to be defined over the field of definition of B_+.*

b) *There are automorphisms indexed by $\tilde{\Omega}$ on the coordinate ring of $E_\alpha(u)$ which are independent of T (but dependent on $u \in G(F)$) such that for any T the action of $\mathrm{Gal}(\bar{F}/F)$ on the coordinate ring is given by $\sigma(x(\sigma^{-1}p)) = (\varphi(\sigma)(x))(p)$ where x belongs to the coordinate ring of $E_\alpha(u)$, and*

$$p \in E_\alpha(u)(\bar{F}),\ \sigma \in \mathrm{Gal}(\bar{F}/F),\ \varphi : \mathrm{Gal}(\bar{F}/F) \to \tilde{\Omega}.$$

(Note that φ depends on both u and T.)

c) *The automorphisms indexed by $\tilde{\Omega}$ on ξ are given by:*
$\sigma_{\alpha'}(\xi) = \xi, \alpha' \neq \alpha, \alpha'$ *simple*
$\sigma_\alpha(\xi) = (\alpha(X)e(\alpha,\beta)w + 1)\xi$
$\sigma_0(\xi) = 1/\zeta\xi, \xi \in F^\times, \zeta$ *depends on u. If K is the splitting field of B_+ then ζ is a norm of an element in $K^\times$ if and only if the line ℓ_α has rational points.*

d) *The automorphisms indexed by $\tilde{\Omega}$ on w are given by:*
$\sigma_{\alpha''}(w) = w/(w(\alpha+\alpha'')\alpha''(X)e(\alpha'',\alpha) + 1) \quad \alpha'' \neq \alpha$
$\sigma_\alpha(w) = w/(w\alpha(X)e(\alpha,\beta) + 1)$
$\sigma_0(w) = we(\alpha,\beta)/(e(\alpha,\beta')(w\alpha(X)e(\alpha,\beta) + 1))$

When $|\tilde{\Omega} : \Omega| \leq 2$ the above gives the the action on generators. The other automorphisms are obtained by composing these in the appropriate manner.

PROOF. (c), (d). By (1.5) the action of a simple root on $x(\alpha')$ is given by $\sigma_{\alpha''}(x(\alpha')) = x(W(\sigma_{\alpha''}), \alpha')$. Thus we may apply ($V$.6.1.$f$). Now $\lambda = x(\alpha)x(\beta)w/x(\gamma)$ so that

$$x(W'',\alpha)x(W'',\beta)\sigma_{\alpha''}(w)/x(\alpha)x(\beta)w = 1$$

or $\sigma_{\alpha''}(w) = w(x(\alpha)/x(W'',\alpha))(x(\beta)/x(W'',\beta))$ where $W'' = W(\sigma_{\alpha''})$. Using ($V$.6.1.$f$) we obtain the results (d) for simple roots.

By (V.6.1.c), $x(\beta)/x(\gamma) = \xi e(\alpha,\beta)$ so that $x(W'',\beta)/x(\beta) = \sigma_{\alpha''}(\xi)/\xi$. Using ($V$.6.1.$f$) once again we obtain the results (c) for simple roots.

Now we turn to the action of σ_0. We write $\xi' = \sigma_0(\xi)$ and $w' = \sigma_0(w)$. By (1.5)

$$\mathbf{ad}\ w_\sigma[a_\sigma(b), B_0^{a_\sigma(n_{w'})})^{\sigma(\nu)}]$$

gives

$$(\mathbf{ad}\ w_\sigma)(B_0^{a_\sigma(\nu)}) = B_0^{\nu'}.$$

Now on $E_\alpha(u)$, $\nu = \exp(\xi X_{-\alpha})$ and $a_\sigma(\exp(\xi X_{-\alpha})) = \exp(x\xi X_{-\alpha})$ for some x. Now apply (V.5.3.a). We obtain $\xi' = 1/\zeta\xi$ some $\zeta \in \bar{F}^\times$ and $(z(W_+,\alpha)/\xi + 1) = (z'(W_+,\alpha)/\xi' + 1)^{-1}$, or by ($V$.2.4)

$$\alpha(X)we(\alpha,\beta) + 1 = (\alpha(X')w'e(\alpha,\beta') + 1)^{-1}.$$

Now $\alpha(X') = -\alpha(X)$ from which the action on w follows immediately.

The equation $w = 0$ defines E_0. Since ζ is independent of w it is enough to verify the properties of ζ in (c) on points of $E_0 \cap E(\ell_\alpha, u)$. This is a projective line isomorphic over F to the line ℓ_α in $(B\backslash G)_u$. We see that for $p \in E_\alpha(u) \cap E_0(F)$ we have

$$\sigma(\xi) = \xi, \sigma|_K = 1$$
$$\sigma(\xi)\xi = \zeta, \sigma|_K \neq 1$$

for $\sigma \in \mathrm{Gal}(\bar{F}/F)$. This shows that $\zeta \in F^\times$ and that ℓ_α has rational points if and only if ζ is a norm in K/F. The same conclusion holds when expressed in terms of the extended Weyl group $\tilde{\Omega}$.

We must also discuss the rationality on the intersection of two divisors which are interchanged for $G = {}^2A_{2n}$. In this case we have

LEMMA 1.7. *Suppose that two subregular divisors E_α and E_β are interchanged by $\mathrm{Gal}(K/F)$ where α and β are adjacent roots and E is the field of definition of E_α. Select coordinates such that B_0 lies at the intersection of the line of type α and the line of type β in $(B\backslash G)_u$. Let B_∞ be any Borel subgroup opposite to B_0. Then the automorphisms indexed by $\tilde{\Omega}$ on w in the coordinate ring of $E_\alpha(u) \cap E_\beta(u)$ are given by*

$\sigma_\alpha(w) = w/(\alpha(X)e(\alpha,\beta)w + 1)$
$\sigma_\beta(w) = w/(\beta(X)e(\beta,\alpha)w + 1)$
$\sigma_{\alpha'}(w) = w, \qquad \alpha' \neq \alpha, \beta$
$\sigma_0(w) = -w.$

PROOF. By $(V.5.1.f)$, the coordinate w is independent of the choice of B_∞. In particular we may calculate the action of $\sigma_{\alpha'}(\alpha'$ simple) on w by restricting the action of $\sigma_{\alpha'}$ on w in the coordinate ring of $E_\alpha(u)$ to $E_\alpha(u) \cap E_\beta(u)$.

Now turn to the action of σ_0. We have (1.5)

$$(a_\sigma(b), B_0^{a_\sigma(n_{w'})})^{a_\sigma(\nu)} = (b', B_0^{n'_w})^{\nu'}.$$

For $W = W_+$ we see that $a_\sigma(\nu) = \nu'$, $a_\sigma(b) = b'$. Thus $x'(\alpha) = x(\beta), x'(\beta) = x(\alpha)$. Furthermore on $E_\alpha \cap E_\beta$ we have $x'(\alpha+\beta) = -x(\alpha+\beta)$. The relations

$$\lambda = x(\alpha)x(\beta)w/x(\alpha+\beta) \text{ and } \lambda = x'(\alpha)x'(\beta)w'/x'(\alpha+\beta)$$

yield

$$w' = (x(\alpha)/x'(\alpha))(x(\beta)/x'(\beta))(x'(\alpha+\beta)/x(\alpha+\beta))w$$

restricting to $E_\alpha \cap E_\beta$ we obtain by $(V.5.1)$ $\sigma_0(w) = w' = -w$.

From (1.6) it is clear that for any σ, $(\sigma^{-1}\xi(\sigma p))\xi^{-1}$ is a rational function of w (provided $\xi = 0$ is defined over F). It appears that the coordinate system breaks down at the finitely many zeros and poles of this rational function. However, we wish to use the coordinates (w, ξ) to fulfill the condition $(V.3.2)$. The following lemma shows that zeros and poles never create a problem because they are not F-rational points.

LEMMA 1.8. *Suppose that ℓ_β is fixed by* $\mathrm{Gal}(\bar{F}/F)$. *Fix* $\sigma \in \mathrm{Gal}(\bar{F}/F)$ *and that $\xi = 0$ is defined over F. The zeros and poles of the rational function of w :*

$$\sigma^{-1}\xi(\sigma p)\xi^{-1}$$

are not F-rational points.

PROOF. For an F-rational points $\sigma p = p$ and $\sigma\xi(\sigma^{-1}p)\xi^{-1}$ is a cocycle of σ with values in $K_X(w)^\times$ where K_X is defined in $(V.2)$. A choice of Cartan subgroup makes K_X into a $\mathrm{Gal}(\bar{F}/F)$-module. We take the cocycle relative to some finite extension K of F and restrict the cocycle to the cyclic group generated by σ. Suppose that the order of σ is ℓ. There is a short exact sequence

$$1 \to K_X^\times \to K_X(w)^\times \to D_0 \to 1$$

where D_0 are the degree zero divisors on $\mathbb{P}^1$, i.e., formal finite sums $\sum n_x x$ with $n_x \in \mathbb{Z}$ and $x \in \mathbb{P}^1$ with $\sum n_x = 0$. This gives a homomorphism

$$H^1(Gal, K_X(w)^\times) \to H^1(Gal, D_0)$$

where D_0 is considered as a $\mathrm{Gal}(K/F)$-module in the obvious way. The cocycle relation becomes

$$(1 + \sigma + \ldots + \sigma^{\ell-1})\sum n_x x = 0.$$

If x_0 is a rational point, $\sigma(x_0) = x_0$ and the cocycle condition becomes

$$\ell n_{x_0} x_0 + (1 + \sigma + \ldots \sigma^{\ell-1})\sum n_x x = 0.$$

This forces $n_{x_0} = 0$.

2. The Characters $\kappa(E_\alpha)$.

The results of this section assume that G is quasi-split. The quasi-split form of G provides an action of $\mathrm{Gal}(\bar{F}/F)$ on the simple roots of G. For each root α there is a field extension F_α of F defined as the smallest extension over which the roots in the orbit of α becomes fixed. F_α is Galois and $F_\alpha = F_{\sigma * \alpha}$ for $\sigma \in \mathrm{Gal}(F_\alpha/F)$. Let Δ' be a set of representatives of the orbits under this action.

The function $e(p)$ can be considered a function of the coordinates $z(W, \alpha)$ for all (W, α). Fix $\Sigma \in \Delta'$ and let

$$e = e(z(W,\alpha) : \alpha \notin \Sigma;\ z(W,\alpha) : \alpha \in \Sigma).$$

And let

$$e' = e(z(W,\alpha) : \alpha \notin \Sigma;\ t(\alpha)z(W,\alpha) : \alpha \in \Sigma)$$

where $t(\alpha) \in F_\alpha$ and $\sigma(t(\alpha)) = t(\sigma_*\alpha)$ for $\sigma \in \mathrm{Gal}(F_\alpha/F)$. Then by [15,5.4] and its generalization in [17] there is a character κ^α of $F_\alpha^\times$ such that $m_\kappa(e') = \kappa^\alpha(t(\alpha))m_\kappa(e)$. With these conventions

$$\kappa^\alpha(t(\alpha)) = \kappa^{\sigma_*\alpha}(t(\sigma_*\alpha))$$

so that if we act on characters of F_α by $\sigma(\theta)(x) = \theta(\sigma^{-1}x)$, $\sigma \in \mathrm{Gal}(F_\alpha/F)$, $x \in F_\alpha^\times$, then $\sigma(\kappa^\alpha) = \kappa^{\sigma*\alpha}$.

We have from (*IV*.1.2)

$$T(W,\alpha)\lambda = z(W,\alpha)x(W,\alpha).$$

On the regular divisor E_0 the coefficients $x(W,\alpha) = x(\alpha)$ are independent of W. Choosing root vectors X_α over F_α such that $\sigma(X_\alpha) = X_{\sigma_*\alpha}$ $\sigma \in \mathrm{Gal}(F_\alpha/F)$ we have $x(\alpha) \in F_\alpha^\times$ and $\sigma(x(\alpha)) = x(\sigma_*\alpha)$. The function $m_\kappa(e)/\prod \kappa^\alpha(\lambda)$ extends to the regular divisor E_0. Products are taken over $\alpha \in \Delta'$ unless indicated otherwise. The restriction to E_0 of this extension equals

$$\prod \kappa^\alpha(1/x(\alpha))m_\kappa(e')$$

where now $e' = e(T(W,\alpha) : \alpha \text{ simple})$. We set $\Delta_\Gamma = m_\kappa(e')$. Also note that by the discussion (*V*.2) of $\kappa(E)$, $\prod \kappa^\alpha = \kappa(E_0)$ which we abbreviate to κ_0. Thus $m_\kappa(e)/\kappa_0(\lambda)$ extends to a regular divisor and equals $\Delta_\Gamma \prod \kappa^\alpha(1/x(\alpha))$. Note that $\prod \kappa^\alpha(1/x(\alpha))$ is none other than $\mu_\kappa(n)$ of [15]. [15] shows that $\mu_\kappa(n)$ and hence $\prod \kappa^\alpha(1/x(\alpha))$ depends on H and not directly on T. This gives the results:

COROLLARY 2.1. *κ^α depends only on H not T.*

COROLLARY 2.2. *Let E be any fundamental divisor defined over F. Then $\kappa(E)$ depends only on H not T. In fact, $\kappa(E) = \prod(\kappa^\alpha)^{e(\alpha)}$ where α runs through representatives in Δ' such that $z(\alpha) = 0$ on E and the Weil divisor defined by the regular function $z(\alpha)$ contains the divisor E with multiplicity $e(\alpha)$.*

PROOF. Select a local coordinate system $\mu_1, \ldots, \mu_n$ so that $\mu_1 = 0$ defines a fundamental divisor E. For every root α, pull the functions $z(\alpha)$ up to Y_Γ and write $z(\alpha) = \mu_1^{e(\alpha)}\xi_\alpha$ where ξ_α is regular and invertible on E. $e(\alpha)$ depends only on the orbit containing α. $e(\alpha)$ is given geometrically as the multiplicity with which E occurs in the Weil divisor determined by $z(\alpha)$. Then

$$\begin{aligned} m_\kappa(e(z(W,\alpha)) &= m_\kappa(e(\mu_1^{e(\alpha)}\xi_\alpha z_1(W,\alpha)) \\ &= \prod \kappa^\alpha(\mu_1)^{e(\alpha)} m_\kappa(e(\xi_\alpha z_1(W,\alpha)). \end{aligned}$$

By the definition of fundamental divisors $z_1(W,\alpha)$ is regular and generically invertible on E. So $m_\kappa(e(\xi_\alpha z_1(W,\alpha))$ extends to a locally constant function on an open set of E. By the discussion (*V*.2), $\kappa(E)$ is the character such that $f/\kappa(E)(\mu_1)$ extends to a locally constant function on an open set of E. Thus we have $\kappa(E) = \prod(\kappa^\alpha)^{e(\alpha)}$.

By corollary 2.2, to determine $\kappa(E)$ for all fundamental divisors it is enough to calculate κ^α for α simple. We observe that if $G \neq {}^2A_{2n}$, Δ' can be selected so that whenever $(\alpha,\beta) \neq 0$ for $\alpha \in \Delta'$ and $\beta \notin \Delta'$ then $F_\alpha = F$. For $G = {}^2A_{2n}$ we let $\Delta' = \{\alpha_1, \ldots, \alpha_n\}$.

PROPOSITION 2.3. *The characters κ^α must have the following form.*

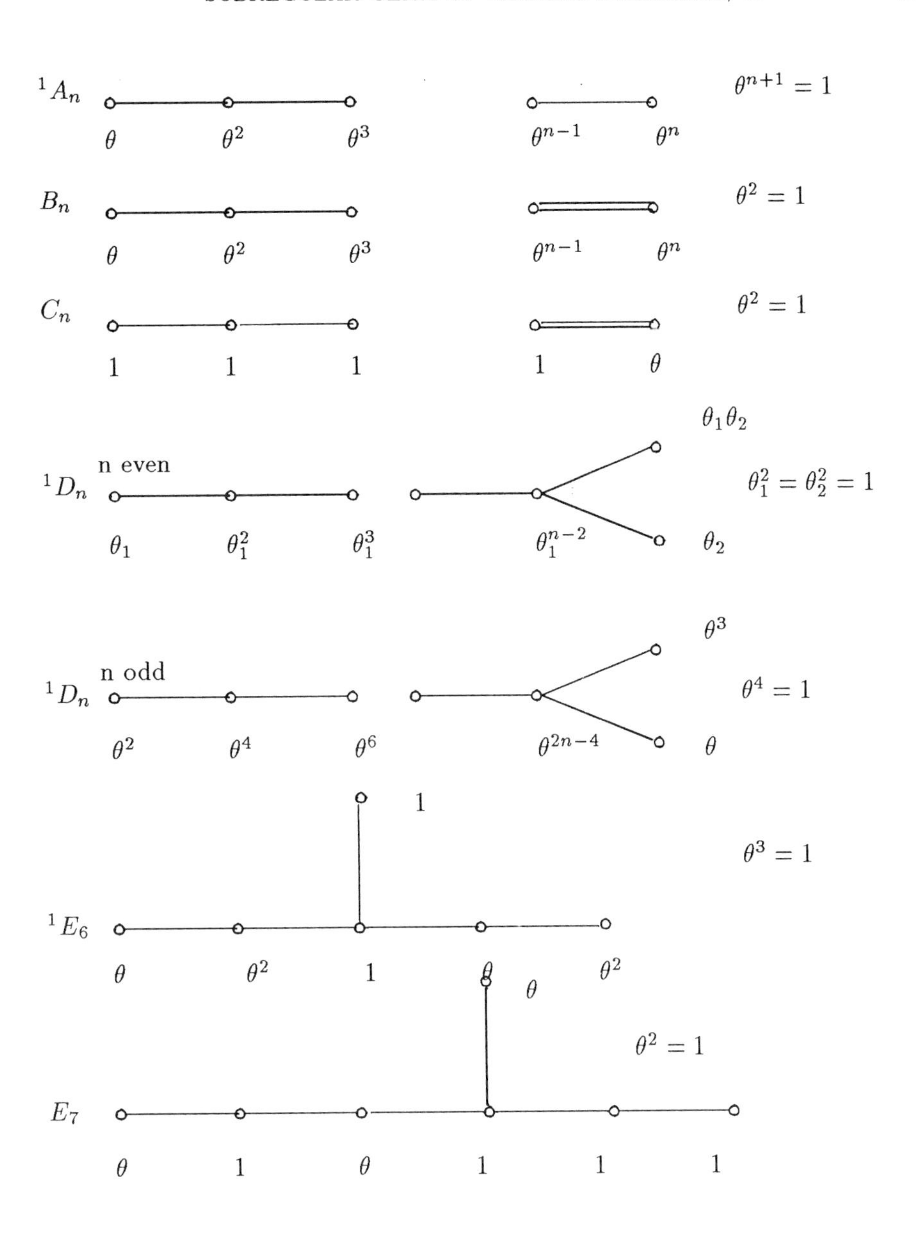

^1A_n
θ
θ^2
θ^3
θ^{n-1}
θ^n
$\theta^{n+1} = 1$
B_n
θ
θ^2
θ^3
θ^{n-1}
θ^n
$\theta^2 = 1$
C_n
1
1
1
1
θ
$\theta^2 = 1$
^1D_n
n even
θ_1
θ_1^2
θ_1^3
θ_1^{n-2}
$\theta_1\theta_2$
θ_2
$\theta_1^2 = \theta_2^2 = 1$
^1D_n
n odd
θ^2
θ^4
θ^6
θ^{2n-4}
θ^3
θ
$\theta^4 = 1$
$^1E_6$
1
θ
θ^2
1
θ
θ^2
$\theta^3 = 1$
E_7
θ
θ
1
θ
1
1
1
$\theta^2 = 1$

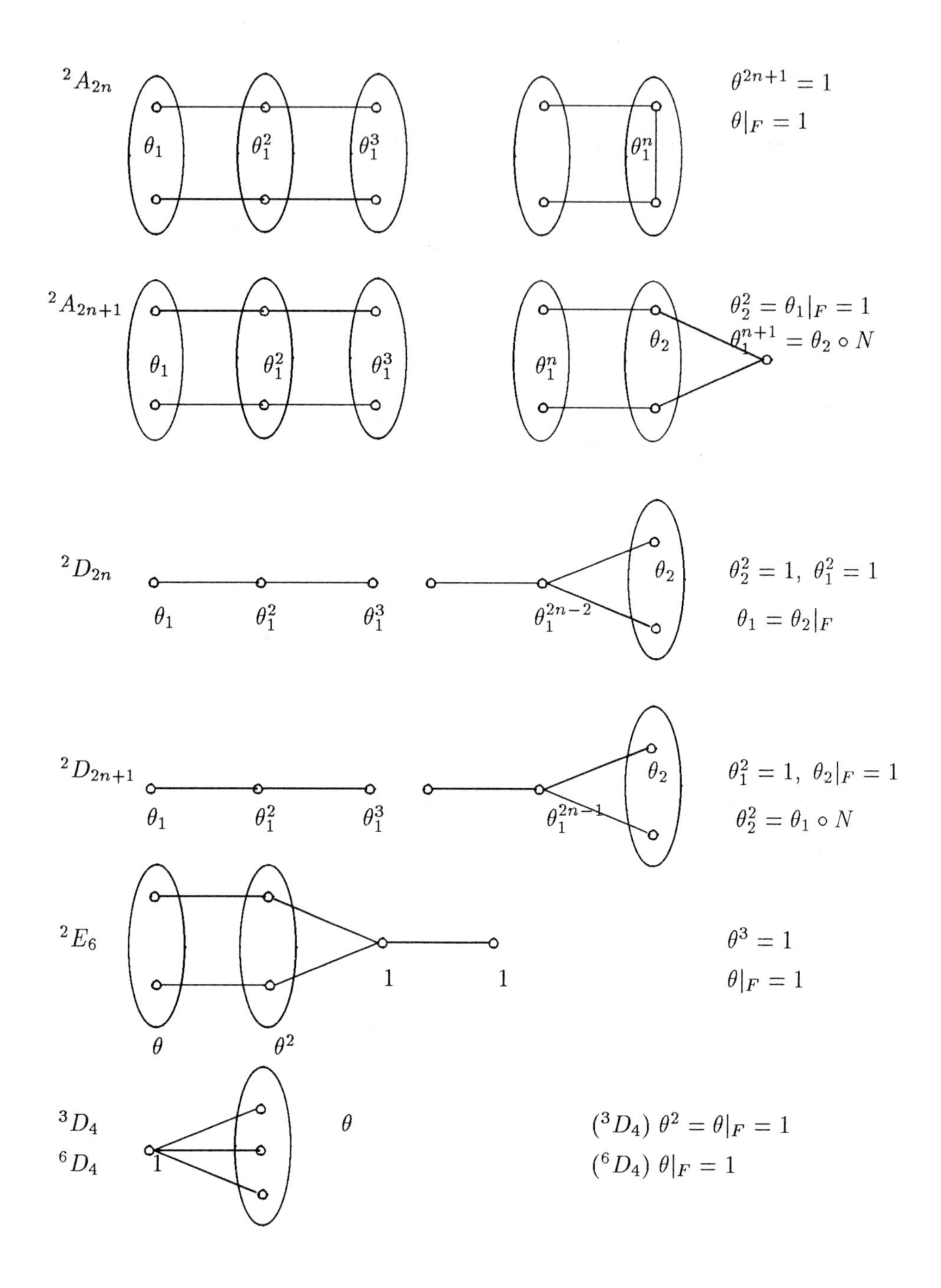

$^2A_{2n}$
θ_1
θ_1^2
θ_1^3
θ_1^n
$\theta^{2n+1} = 1$
$\theta|_F = 1$
$^2A_{2n+1}$
θ_1
θ_1^2
θ_1^3
θ_1^n
θ_2
$\theta_2^2 = \theta_1|_F = 1$
$\theta_1^{n+1} = \theta_2 \circ N$
$^2D_{2n}$
θ_1
θ_1^2
θ_1^3
θ_1^{2n-2}
θ_2
$\theta_2^2 = 1,\ \theta_1^2 = 1$
$\theta_1 = \theta_2|_F$
$^2D_{2n+1}$
θ_1
θ_1^2
θ_1^3
θ_1^{2n-1}
θ_2
$\theta_1^2 = 1,\ \theta_2|_F = 1$
$\theta_2^2 = \theta_1 \circ N$
$^2E_6$
1
1
θ
θ^2
$\theta^3 = 1$
$\theta|_F = 1$
$^3D_4$
$^6D_4$
1
θ
$(^3D_4)\ \theta^2 = \theta|_F = 1$
$(^6D_4)\ \theta|_F = 1$

The character κ^α is a character on $F_\alpha^\times$. N denotes the norm map from K to F. $|_F$ denotes the restriction to F. The characters are trivial on G_2, F_4, E_8.

PROOF. Consider a coordinate patch $S(B_\infty, B_0)$ with Cartan subgroup $T_0 = B_\infty \cap B_0$ and Borel subgroup B_0 defined over F. Then on $Y^0(B_\infty, B_0)$, $\mathbb{B}(W)^h = B_0^\omega$, $W = W(\omega)$ for some $h \in G(\bar{F})$. For $t' \in T_0(F)$, let e be given by $(B_0^{\omega n})^\nu$ and e' by $(B_0^{\omega n})^{\nu t} = (B_0^{\omega n'})^{\nu'}$. Then

$$\begin{aligned} m_\kappa(e) &= \kappa(\sigma(h)\sigma(n)\sigma(\nu)\nu^{-1}n^{-1}h^{-1}) \\ &= \kappa(\sigma(h)\sigma(n)\sigma(\nu)\sigma(t')t'^{-1}\nu^{-1}n^{-1}h^{-1}) \\ &= m_\kappa(c'). \end{aligned}$$

If $e = e(z(W,\alpha))$, $e' = e(z'(W,\alpha))$, $n' = adt'^{-1}(n)$, and

$$z'(W,\alpha) = \alpha(t')z(W,\alpha)$$

then

$$m_\kappa(e) = \prod \kappa^\alpha(\alpha(t'))m_\kappa(e'),$$

$\alpha \in \Delta'$.

Thus $\prod \kappa^\alpha(\alpha(t')) = 1$ for all $t' \in T_0(F)$. If β^v is the coroot of $\beta \in \Delta'$ and if $z \in F_\beta^\times$ then

$$\prod \sigma(z)^{\sigma(\beta^v)} \in T_0(F)$$

(product taken over $\sigma \in \mathrm{Gal}(F_\beta/F)$). We obtain

$$\prod \kappa^\alpha(\alpha(\prod \sigma(z)^{\sigma(\beta^v)})) = \prod \kappa^\alpha(\sigma(z))^{<\alpha,\sigma(\beta^v)>} = 1.$$

The second product extends over $\alpha \in \Delta'$ and $\sigma \in \mathrm{Gal}(F_\beta/F)$. By the choice of Δ' : $<\alpha,\sigma(\beta^v)> = 0$ if $\sigma \neq 1$ in $\mathrm{Gal}(F_\beta/F)$ (for $G \neq {}^2A_{2n}$). We break the product into two pieces. The first is

$$\prod \kappa^\alpha|_{F_\beta}(z)^{<\alpha,\beta^v>}$$

where the product extends over all $\alpha \in \Delta'$ such that $F_\alpha \neq F$. The second is

$$\prod \kappa^\alpha(Nz)^{<\alpha,\beta^v>}$$

where N is the norm map from F_β to F and the product here extends over all $\alpha \in \Delta'$ such that $F_\alpha = F$. The result is now a short calculation carried out but substituting successively all coroots in for β^v.

COROLLARY 2.4. *Let G be split. $\kappa(E_0) =^{(def)} \kappa_0$ is given by:*
a) If $G = A_{2n}, G_2, F_4, E_6, E_8$ then κ_0 is the trivial character.
b) For $A_{2n+1}\kappa_0 = \theta^p$ where $p = n+1$ and $\theta^{2p} = 1$
c) For $B_n, \kappa_0 = \theta^{n(n+1)/2}$ where $\theta^2 = 1$
d) For $C_n, \kappa_0 = \theta$ where $\theta^2 = 1$
e) For $D_{2n}, \kappa_0 = \theta_1^n$ where $\theta_1^2 = 1$
f) For $D_{2n+1}, \kappa_0 = \theta^{2n}$ where $\theta^4 = 1$
g) For $E_7, \kappa_0 = \theta$ where $\theta^2 = 1$.

PROOF. $\kappa_0 = \prod \kappa^\alpha|_F$.

3. $m_\kappa(e)$ and Vanishing of Integrals.

LEMMA 3.1. *Fix a surface $E(\ell_\alpha, u)$. Suppose that for all $\beta \neq \alpha$ either $E(\ell_\alpha, u) \cap E_\beta$ has no F-rational points or $\kappa(E_\beta) \neq \kappa(E_\alpha)$. Then the principal value integral can be computed on any variety birational to $E_\alpha(u)^0$ which is biregular on $E(\ell_\alpha, u) \cap E_0$. In particular, the principal value integral depends only on the structure of the divisor on its intersection with Y''.*

PROOF. Any birational map on a surface can be factored by successively blowing up and down at points. It therefore suffices to prove that the principal value integral is unaffected by blowing up. We write the form locally as

$$\gamma \mu_1^{b_1-1} \mu_2^{b_2-1} d\mu_1 \wedge d\mu_2, |\gamma| \text{ constant}.$$

If b_i is not equal to one then $\mu_i = 0$ defines the intersection of $E(\ell_\alpha, u)$ with a divisor E' and $b_i = b_\alpha(E')$. The constants $b_\alpha(E)$ are related to the constants of the Igusa data by the relation (V.2) $b_\alpha(E) = b(E) - 2a(E)$. In particular, $b_\alpha(E_0) = -1$, $b_\alpha(E_\beta) = 0$ (β simple), and $b_\alpha(E) \geq 1$ otherwise. Blowing up at $\mu_1 = \mu_2 = 0$ creates a divisor E with $b_\alpha(E) = \sum b_i$. The conditions of the lemma insure that the principal value integral on $E(\ell_\alpha, u)$ is well defined (that is, it is not necessary to resort to the definitions of (V.3)). [16] guarantees that blowing up does not alter principal value integrals provided $b_\alpha(E) \neq 0$. If $\sum b_i = 0$ then permuting coordinates if necessary either $b_1 = b_2 = 0$ or $b_1 = -1, b_2 = +1$. The first possibility never arises because three lines of the Dynkin curve $(B\backslash G)_u$ never intersect. The second possibility is excluded by the condition that the rational map be biregular on E_0.

PROPOSITION 3.2. *Assume that ℓ_α intersects one line ℓ_β and that ℓ_α is fixed by $\mathrm{Gal}(\bar{F}/F)$.*

a) If $\kappa(E_\alpha) \neq \kappa(E_\beta)$ then $E(\ell_\alpha, u)$ makes no contribution to the subregular germ.

b) If $\kappa(E_\alpha) = \kappa(E_\beta)$ then $\kappa(E_0) = \kappa(E_\alpha)$ and $m_\kappa(e)/\kappa_0(\lambda)$ restricted to $E_\alpha(u)$ depends only on $u \in G(F)$. If we select coordinates as in (V.6.2) with T_0 defined over F, it equals

$$\prod \kappa^{\alpha'}(1/x(\alpha'))\Delta_\Gamma.$$

PROOF. Note that the hypothesis implies that G is quasi-split. I claim that $z(\alpha)$ is defined over F. $\lambda = x(\alpha)z(\alpha)$, $\lambda = x(W, \alpha)\sigma^{-1}(z(\alpha)(\sigma p))$. By (V.6.2) we see that $x(W, \alpha)/x(\alpha) = 1$ on E_α so $z(\alpha)$ is defined over F.

$$\begin{aligned} z(W, \alpha') &= T(W, \alpha')\lambda/x(W, \alpha') \\ &= (x(\alpha')/x(W, \alpha'))T(W, \alpha')(\lambda/x(\alpha')) : \alpha' \neq \alpha \\ &= (x(\alpha)/x(W, \alpha))T(W, \alpha)z(\alpha) : \alpha' = \alpha. \end{aligned}$$

Again by (V.6.2), $x(\alpha')/x(W, \alpha') = 1$ on $E_\alpha(u)$.

$$m_\kappa(e)/\kappa(E_\alpha)(\lambda)|_{E_\alpha} = \prod \kappa^{\alpha'}(1/x(\alpha'))\kappa^\alpha(z(\alpha))\Delta_\Gamma.$$

Using arguments as in (V.2) we find that the differential form is

$$(d\lambda/\lambda^2) \wedge dx(\alpha) \wedge \ldots =$$

$$dz(\alpha)/z(\alpha)^2 \wedge dx(\alpha)/dx(\alpha) \wedge \ldots$$

and is $dz(\alpha)/z(\alpha)^2 \wedge d\xi$ on the fibre up to a scalar independent of T.

We have a morphism over F from an open set of $E(\ell_\alpha, u)$ to $\mathbb{P}^1$ given by $(\xi, z(\alpha)) \rightarrow (z(\alpha)) \in \mathbb{P}^1$. We extend this morphism by blowing up a finite number of points of $E(\ell_\alpha, u)$. If blowing up is necessary at points of $E_0 \cap E(\ell_\alpha, u)$ we must check that no exceptional divisors E with $b_\alpha(E) = 0$ are introduced. By the calculations of (V.5.3)

$$(a_w + \xi)/\xi = \xi''/(a''_w + \xi''), \xi'' = 1/\xi\zeta$$

where ξ'' and w'' are canonical coordinate on $E(\ell_\alpha, u) \cap U(\alpha, \beta)$. Now by ($V$.2.4) $(a''_w + \xi'')/\xi''$ for $W = W(\sigma_\alpha)$ equals $\alpha(X)e(\alpha, \beta)w'' + 1$.

$$(a_w + \xi)/\xi = (T(W, \alpha)z(\alpha) + \xi)/\xi = T(W, \alpha)z(\alpha)/\xi + 1,$$

or

$$z(\alpha) = -\alpha(X)e(\alpha, \beta)w''/(\alpha(X)e(\alpha, \beta)w'' + 1)\zeta\xi''.$$

This morphism does not extend to $w'' = \xi'' = 0$. This defines the point of intersection of $E(\ell_\alpha, u)$ with $E_0 \cap E_\beta$. Blowing up creates a divisor E with $b_\alpha(E) = b_\alpha(E_0) + b_\alpha(E_\beta) = -1 + 0 = -1 \neq 0$. [16] tells us this does not affect the principal value integral. It is easy to see that by blowing up once the morphism extends to points over $w'' = \xi'' = 0$.

Now with the morphism $E_\alpha(u) \rightarrow \mathbb{P}^1$ we integrate over the fibre using [16]

$$\int |d\xi| = 0.$$

LEMMA 3.3. *Suppose that ℓ_α intersects two lines ℓ_β and $\ell_{\beta'}$. Suppose also that* $\mathrm{Gal}(\bar{F}/F)$ *fixes these two points of intersection. Pick canonical coordinates* (w, ξ) *on* $U(\alpha, \beta)$. *Fix* $u \in G(F)$.

a) *If* $\kappa^\alpha \neq \kappa^\beta$ *then* $E_\alpha(u)$ *makes no contribution to the subregular germ.*

b) *If* $\kappa^\alpha = \kappa^\beta$ *then* $\kappa(E_0) = \kappa(E_\alpha)$, $\kappa^\alpha = \kappa^\beta = \kappa^{\beta'} = 1$, *and*

$$m_\kappa(e)/\kappa_0(\lambda)|_{E_\alpha} = \Delta_\Gamma \prod \kappa^{\alpha'}(1/x(\alpha')).$$

REMARK 3.4. The formulas in (b) and (3.2.b) effectively allow us to ignore the transfer factor on $E(\ell_\alpha, u)$ for all but possibly one surface (for a simple group) when considering the question of the transfer of the subregular germ of orbital integrals. If ℓ_α intersects three lines or if ℓ_α intersects two lines that are interchanged by $\mathrm{Gal}(\bar{F}/F)$, then $m_\kappa(e)$ must be computed using (I.5).

PROOF. The hypothesis forces G to be quasi-split. Assume that $\kappa^\alpha \neq \kappa^\beta$. Then on an open set in $E_\alpha(u)$ we define a morphism over F to $\mathbb{P}^1$ by $(w,\xi)\to w$. We extend this morphism by blowing up at a finite number of points over $E_\alpha(u)$. By (3.1), we must verify that if blowing up is required at points over E_0 that no divisors E are introduced with $b_\alpha(E) = 0$. We express the morphism in terms of canonical coordinates (w'',ξ'') on $U(\alpha,\beta')$. By (V.5.3),

$$e(\alpha,\beta)w = -e(\alpha,\beta')w''/(\alpha(X)e(\alpha,\beta')w''+1).$$

From this expression it is clear that this morphism extends to points of E_0 ($w''=0$) on the coordinate patch $U(\alpha,\beta')$. The patches $U(\alpha,\beta)$ and $U(\alpha,\beta')$ cover $E_0 \cap E_\alpha(u)$ so that the morphism extends without difficulty.

We select F-coordinates. By (1.6) $\sigma\xi(\sigma^{-1}p)\xi^{-1} = a_\sigma(w)$ where for fixed σ this is a rational function of w. Define an action σ_* of $\mathrm{Gal}(\bar F/F)$ on (w,ξ) by

$$\sigma_*(w(p)) = \sigma(w(\sigma^{-1}p))$$
$$\sigma_*(\xi(p)) = \sigma(\xi(\sigma^{-1}p)).$$

With respect to this action $a_\sigma(w)$ is a cocycle of $\mathrm{Gal}(\bar F/F)$ with coefficients in $K_X(w)^\times$. By Hilbert's theorem 90 there is an element $b(w)\in K_X(w)^\times$ such that $a_\sigma(w) = \sigma_*(b)b^{-1}$. This gives an F-coordinate $\xi' = b^{-1}\xi$ (away from the zeros and poles of b).

Next we compute $m_\kappa(e)/\kappa(E_\alpha)(\lambda)$ on E_α.

$$\begin{aligned} z(W,\alpha') &= T(W,\alpha')\lambda/x(W,\alpha') \\ &= (x(\alpha')/x(W,\alpha'))T(W,\alpha')(\lambda/x(\alpha')) : \alpha'\neq\alpha \\ &= (x(\alpha)/x(W,\alpha))T(W,\alpha)z(\alpha) : \alpha' = \alpha. \end{aligned}$$

The dependence on ξ' of the right hand side of this equation is through

$$x(\beta)|_{E_\alpha} = \xi e(\alpha,\beta)x(\alpha+\beta)$$

(V.6.1.c)) and $z(\alpha) = z(W_+,\alpha)/T(W_+,\alpha) = e(\alpha,\beta)\xi w\alpha(X)/T(W_+,\alpha)$. This uses (V.6.1.$d$).

It follows that $m_\kappa(e)/\kappa(E_\alpha)(\lambda)$ restricted to E_α equals

$$(*) \qquad \prod \kappa^{\alpha'}(1/x(\alpha'))\kappa^\alpha(\xi')\kappa^\beta(1/\xi')m_\kappa(e')$$

where e' depends on w not ξ'. When $\kappa^\alpha\neq\kappa^\beta$ we integrate over ξ' and use [16]

$$\int \kappa^\alpha/\kappa^\beta(\xi')|d\xi'/d\xi'| = 0.$$

(b) Referring to (2.3), we observe that the characters of α,β,β' are in geometric progression: $\kappa^\beta = \theta^{i-1}, \kappa^\alpha = \theta^i, \kappa^{\beta'} = \theta^{i+1}$ for some θ and i. Thus $\kappa^\alpha = \kappa^\beta$ implies that $\kappa^\alpha = \kappa^\beta = \kappa^{\beta'} = 1$. We also observe that there is a chain of lines $\ell_{\alpha_1},\dots,\ell_{\alpha_k}$ with $\ell_{\alpha_{k-1}} = \ell_\alpha, \ell_{\alpha_k} = \ell_\beta, \kappa_{\alpha_i} = 1$. Lemma (3.2) shows that $m_\kappa(e)$ is constant on $E_{\alpha_1}(u)$. The expression $(*)$ with $\kappa^\alpha/\kappa^\beta = 1$ shows that $E_\alpha(u)$ is independent of ξ. This is also true for canonical coordinates on $E_{\alpha_2}(u), \ldots, E_{\alpha_k}(u)$. By induction we may assume $E_{\alpha_{j-1}}(u)$ is constant so that $E_{\alpha_j}(u)$ is constant on $E_{\alpha_j}(u) \cap E_{\alpha_{j-1}}(u)$, i.e. for $\xi = 0$. But since $m_\kappa(e)$ is independent of ξ, $m_\kappa(e)$ must then be constant on $E_{\alpha_j}(u)$. Thus finally $m_\kappa(e)$ is constant on $E_\alpha(u)$.

Since $m_\kappa(e)$ is constant on $E_\alpha(u)$, the value of $m_\kappa(e)$ equals the value of $m_\kappa(e)$ for $w = 0, i.e.$ on E_0. On E_0 we have

$$z(W, \alpha') = (x(\alpha')/x(W, \alpha'))T(W, \alpha')(\lambda/x(\alpha')).$$

For a regular element $x(\alpha') = x(W, \alpha')$ for all W, and

$$m_\kappa(e)/\kappa_0(\lambda) = \prod \kappa^{\alpha'}(1/x(\alpha'))\Delta_\Gamma.$$

This definition extends to points in the intersection of E_0 with E_α because the characters $\kappa^\alpha, \kappa^\beta, \kappa^{\beta'}$ are trivial.

VII. APPLICATIONS TO ENDOSCOPIC GROUPS

This chapter discusses applications of the formula for subregular germs to endoscopic groups. We will begin by listing the cuspidal endoscopic groups.

1. Endoscopic Groups

Basic facts about endoscopic groups will be assumed. For definitions see [14]. For our purposes, the most important properties of endoscopic groups H that will be used are:

1) The identity component of the dual ${}^L H^0$ of H is the connected centralizer of a semisimple element $s \in {}^L G^0$.
2) There is a homomorphism ρ from $\mathrm{Gal}(\bar{F}/F)$ to the group of outer automorphisms of ${}^L H^0$ which factors through $Cent(s, {}^L G)/{}^L H^0$.

Following Arthur an endoscopic group is said to be *cuspidal* if there is no proper parabolic subgroup of ${}^L G$ containing ${}^L H$. This chapter will ignore the problem of embeddings of L-groups $\xi : {}^L H \to {}^L G$. For our purposes (1) and (2) may be taken as defining properties of endoscopy. In particular we are not asserting the existence of $\xi : {}^L H \to {}^L G$.

1.1. The groups ${}^1A_n$ and ${}^2A_n$.

We do no compute all the endoscopic groups for groups of type A_n. Instead, for the transfer of the subregular germ, we will appeal to the following Proposition

PROPOSITION 1.1. *Let G be a group of type A_n and let H be an endoscopic group of G. Let K_G and K_H be the smallest field extensions of F over which G and H are inner forms of a split group. If $K_G \neq K_H$ then none of the subregular unipotent classes of H are defined over F.*

PROOF. We may assume G is a form of $SL(n)$. First suppose that $G = SL(n)$. The endoscopic groups of $SL(n)$ are given in [14]. The cuspidal ones are of the form ${}^\ell(A_{r-1} \times \ldots \times A_{r-1})$, $\ell r = n$. They have no subregular class over F unless $\ell = 0$. If $G = SU(n)$, then $H \times Spec(K_G)$ is an endoscopic group of $G \times Spec(K_G) \tilde{\to} SL(n)$ as groups over K_G. Again they have no subregular class over K_G (and hence F) unless $\ell = 0$.

By this result it suffices to compute the endoscopic groups of $SU(n)$ which have the same splitting field K as $SU(n)$. Suppose H is defined by

$$\hat{s} = (s_1 I_{a_1}, \ldots s_k I_{a_k})$$

where $\hat{s} \in GL(n, \mathbb{C})$ maps to s in ${}^LG^0$. Let σ be the nontrivial automorphism of $({}^LG^0, {}^LB^0, {}^LT^0, \{Y_\alpha\})$. The conditions defining endoscopic groups require that $w(\sigma(s)) = s$ for some w in the normalizer of ${}^LT^0$. Or $w\sigma(\hat{s}) = \lambda\hat{s}$ for some $\lambda \in \mathbb{C}^\times$. Now if $\tau^2 = \lambda I_n$ then $w\sigma(\tau)\tau = \tau^{-1}\tau = 1$. So $w\sigma(\hat{s}\tau) = \hat{s}\tau$. So by adjusting the choice of $\hat{s}$ mapping to s if necessary we may assume $w\sigma(\hat{s}) = \hat{s}$. This means that up to isogeny H is an endoscopic group of $U(n)$ (cf. section 2). So without loss of generality we take $G = U(n), \hat{s} = s, {}^LG^0 = GL(n, \mathbb{C})$.

Now $\sigma(s) = (s_k^{-1}I_{a_k}, \ldots, s_1^{-1}I_{a_1})$ and the Weyl group acts as permutations; thus for every i there is a j such that $s_i = s_j^{-1}$. Replacing s by $w'(s)$ for some w' in the normalizer of ${}^LT^0$ we have

$$s = (s_1 I_{a_1}, \ldots, s_p I_{a_p}, I_r, -I_t, s_p^{-1} I_{a_p}, \ldots, s_1^{-1} I_{a_1}).$$

The endoscopic group is not cuspidal unless $p = 0$ which we now assume so $s = (I_r, -I_t)$.

Suppose that the rank is even. Now $r + t = 2n + 1$ so exactly one of r and t is odd. By replacing s by $-s$ if necessary we may assume that $t = 2k$ and $r = 2m + 1, k + m = n$. Then again replacing s by $w''(s)$, w'' in the normalizer of ${}^LT^0$, we may assume $s = (I_k, -I_{2m+1}, I_k)$. With this choice of s the condition $w(\sigma(s)) = s$ together with the condition that $w\sigma$ act as outer automorphisms forces $w = 1$. (Here and elsewhere we identify the group of outer automorphisms with automorphisms that fix a given Borel subgroup, Cartan subgroup, and root vectors.) Thus $H = U(2m + 1) \times U(2k)$.

Now consider $G = U(2n), s = (I_r, -I_t)$. If both $r = 2k$ and $t = 2j$ are even we may assume that $s = (I_k, -I_{2j}, I_k)$ and that $H = U(2k) \times U(2j)$, $j + k = n$. If both $r = 2k + 1$ and $t = 2j + 1$ are odd we may assume that s is $(I_k, -I_j, 1, -1, -I_j, I_k)$ and that w is the simple reflection corresponding to the simple root α fixed by σ. We conclude that the cuspidal endoscopic groups of $U(n)$ are $U(j) \times U(n - j)$.

1.2. Type B_n.

The identity component of the L-group of the simply connected semisimple group of type B_n is ${}^LG^0 = PSp(2n)$. Suppose $\hat{s} \in Sp(2n)$ lies over $s \in PSp(2n)$ defining the endoscopic group. Without loss of generality we may take $\hat{s}$ to be

$$\hat{s} = (s_1 I_{a_1}, -s_1 I_{a_2}, \ldots, -s_p I_{a_p}, iI_q, I_r, -I_{2t}, I_r, -iI_q, \ldots, s_1^{-1} I_{a_1}).$$

The group is not cuspidal unless $p = 0$ which we now assume; so

$$\hat{s} = (iI_q, I_r, -I_{2t}, I_r, -iI_q)$$

If ρ is trivial this is not cuspidal unless $q = 0$; so $\hat{s} = (I_r, -I_{2t}, I_r)$, ${}^LH^0 = C_r \times C_t$, $r + t = n$ and $H = B_r \times B_t, r + t = n$. If ρ is nontrivial then $r = t$ and ${}^LH^0 = A_{q-1} \times C_r \times C_r$ and $H = {}^2A_{q-1} \times {}^2(B_r \times B_r), q + 2r = n$.

1.3. Type C_n.

The connected component of the L-group of the simply connected semi-simple group of type C_n is ${}^LG^0 = SO(2n + 1)$. Without loss of generality, select

$$s = (s_1 I_{a_1}, s_2 I_{a_2}, \ldots, s_p I_{a_p}, -I_m, I_{2r+1}, -I_m, \ldots, s_1^{-1} I_{a_1}).$$

This is not cuspidal unless $p = 0$ which we now assume. So

$$s = (-I_m, I_{2r+1}, -I_m)$$

${}^LH^0 = D_m \times B_r, m + r = n$. $H = C_r \times D_m$ if ρ is trivial and $C_r \times {}^2D_m$ if ρ is non-trivial $(m + r = n)$.

1.4. Type D_n.

The connected component of the L-group of the simply connected semisimple group of type D_n is ${}^LG^0 = PSO(2n)$. We may assume $\hat{s} \in SO(2n)$ to be

$$(*) \qquad \hat{s} = (s_1 I_{a_1}, -s_1 I_{a_1'}, \ldots, -s_p I_{a_p'}, iI_q, I_r, -I_{2t}, I_r, -iI_q, \ldots, s_1^{-1} I_{a_1}).$$

There is some difficulty if $r = t = 0$ for the Weyl group of D_n allows permutations of coordinates but only an even number of sign changes, so that $\hat{s}$ cannot always be brought precisely into this form. But it can be brought into this form by the extended Weyl group. Thus it is possible for there to be two inequivalent endoscopic groups which are isomorphic as reductive groups.

Ignoring this difficulty, we find that the group H is not cuspidal unless $p = 0$ which we now assume. So $\hat{s} = (iI_q, I_r, -I_{2t}, I_r, -iI_q)$. If ρ is non-trivial on $ker(SO(2n) \to PSO(2n))$ then $r = t$ and ${}^LH^0 = A_{q-1} \times D_r \times D_r$. So $H = {}^2A_{q-1} \times {}^2({}^?D_r \times {}^?D_r)$ where $q + 2r = n$. The superscript ? indicates that various quasi-split forms are possible. If ρ is trivial on $ker(SO(2n) \to PSO(2n))$ then H is not cuspidal unless $q = 0$ so that $\hat{s} = (I_r, -I_{2t}, I_r)$ and $H = {}^?D_r \times {}^?D_t$. Again various quasi-split forms are possible.

Allowing D_n to be quasi-split and split over a non-trivial quadratic extension K of F allows little new. The outer automorphism acts on ${}^LH^0$ by an outer automorphism of the factor D_t. If $t = 0$, it acts by an outer automorphism of the factor D_r. If $r = t = 0$ and q is odd we obtain the endoscopic group ${}^2A_{q-1}$. If $r = t = 0$ and q is even then ρ cannot fix s and there is no endoscopic group.

1.5. Type G_2.

Let the roots of G_2 be

$$\pm\alpha, \pm\beta, \pm(\alpha + \beta), \pm(2\alpha + \beta), \pm(3\alpha + \beta), \pm(3\alpha + 2\beta).$$

Suppose that ${}^LH^0$ contains a short root. By equivalence we may assume that it is α. There must be another positive root if H is cuspidal. α together with any positive root other than $3\alpha + 2\beta$ generate all the roots of G_2. So we can take the roots to be α and $3\alpha + 2\beta$ and $H = A_1 \times A_1$. This leaves the case where all the roots of H are long roots. If this is to be a cuspidal group there must be at least two positive roots. These will generate all the long roots. We obtain $H = A_2$.

1.6. Other exceptional groups.

We will not compute these. It should be pointed out however that most of the cuspidal endoscopic groups can be deduced directly from [4] where primitive subalgebras of the exceptional groups are computed.

LEMMA 1.6. *Let $G = F_4, E_s$, $s = 6, 7, 8$. Then the centralizer of a semisimple element in G stabilizes one of the following subalgebras of G.*

Algebra	*Primitive Subalgebras*
E_8	$A_1 \oplus E_7$, A_1^8, $A_2 \oplus E_6$, A_2^4, A_4^2, D_4^2, D_8, A_8, T^8
E_7	$A_1 \oplus D_6$, $A_1^3 \oplus D_4$, A_1^7, $A_2 \oplus A_5$, $A_2^3 \oplus T^1$, A_7, $E_6 \oplus T^1$, T^7
E_6	$A_1 \oplus A_5$, A_2^3, $D_4 \oplus T^2$, $D_5 \oplus T^1$, T^6
F_4	$A_1 \oplus C_3$, A_2^2, B_4, D_4.

T^k denotes the center of the subalgebra, where k is the dimension of that center.

PROOF. This is proved in [4]. The algebra $A_2 \oplus D_5$ listed there as a subalgebra of E_7 is apparently a misprint for the subalgebra $A_2 \oplus A_5$.

2. Characters, Centers, and Endoscopic Groups

The next lemma shows that we do not lose any endoscopic groups by passing to the simply connected cover of the derived group.

LEMMA 2.1. *Let G be a reductive group. Let G_s be a cover of the derived group of G. Let H be an endoscopic group of G. Then there is an endoscopic group H_s of G_s and an isogeny $H_s \to H$.*

PROOF. We have a morphism $G_s \to^{\varphi} G$. Fixing a Borel subgroup B and Cartan subgroup T in G fixes B_s and T_s in G_s. Let $\tilde{K}$ be the L-group ${}^L(G_s)$, and let tildes denote quantities in $\tilde{K}$ corresponding to quantities in ${}^LG$. ${}^LG^0$ is a reductive group whose derived group is a cover of ${}^L(G_s)^0$. Thus we have a surjection ${}^L\varphi^0 : {}^LG^0 \to \tilde{K}^0$ which factors through ${}^L(G_{der})^0$. Since φ is defined over F, ${}^L\varphi^0$ extends to ${}^L\varphi : {}^LG \to \tilde{K}$ [2]. Let $\tilde{x} = {}^L\varphi(x)\ \forall\ x \in {}^LG$. The image of $Cent(s, {}^LG)$ under ${}^L\varphi$ lies in $Cent(\tilde{s}, \tilde{K})$ because $xsx^{-1}s^{-1} = 1$ implies $\tilde{x}\tilde{s}\tilde{x}^{-1}\tilde{s}^{-1} = 1$. Similarly

$$ {}^L\varphi^0(Cent(s, {}^LG^0)) \subseteq Cent(\tilde{s}, \tilde{K}^0) $$

and

$$ {}^L\varphi^0(Cent(s, {}^LG^0)^0) \subseteq Cent(\tilde{s}, \tilde{K}^0)^0. $$

I claim that this last inclusion is actually an equality:

$$ {}^L\varphi^0(Cent(s, {}^LG^0)^0) = Cent(\tilde{s}, \tilde{K}^0)^0. $$

Let

$$ K = ({}^LG^0)_{der} \cap ({}^L\varphi^0)^{-1}(Cent(\tilde{s}, \tilde{K}^0)^0). $$

Then ${}^L\varphi^0(K) = Cent(\tilde{s}, \tilde{K}^0)^0$, so ${}^L\varphi^0(K^0) = Cent(\tilde{s}, \tilde{K}^0)^0$. Now we have a morphism of varieties

$$ K^0 \to ker(({}^LG^0)_{der} \to \tilde{K}^0) \text{ given by} $$

$$ x \to xsx^{-1}s^{-1}. $$

But *ker* is discrete and K^0 is connected and $1 \in K^0$ is sent to $1 \in ker$ so the image $\xi(K^0)$ is 1. That is, $xsx^{-1}s^{-1} = 1$ for all $x \in K^0$. So $K^0 \subseteq Cent(s, {}^LG^0)^0$ and $Cent(\tilde{s}, \tilde{K}^0)^0 = {}^L\varphi^0(K^0) \subseteq {}^L\varphi^0(Cent(s, {}^LG^0)^0)$ proving the equality. Notice too that the kernel is central.

We have a homomorphism

$$\rho : (\mathrm{Gal}(\bar{F}/F) \rightarrow Aut({}^LH^0, {}^LB^0_H, {}^LT^0_H, \{Y_\alpha\}).$$

Let $\tilde{\rho}$ be given by $\tilde{\rho} = {}^L\varphi \circ \rho$. If $\rho(\sigma)$ is given by $\mathbf{ad}n(w) : n(w) \in {}^LG$ then ${}^L\varphi(\rho(\sigma)) = \tilde{\rho}(\sigma)$ is given by $\tilde{n}(w) = {}^L\varphi(n(w)) \in \tilde{K}$. So $\tilde{\rho}$ satisfies the conditions of [14] provided ρ does.

Thus we obtain endoscopic groups H and H_s corresponding to G and G_s respectively. Since we have a surjection ${}^LH^0 \rightarrow {}^L\tilde{H}^0$ with central kernel we obtain a dual morphism $H_s \rightarrow H$ again with central kernel [2].

We have the simple but useful lemma:

LEMMA 2.2. *Let G be a reductive group and let Z_1 be a subgroup over F in the center of G. A necessary and sufficient condition for H to descend to an endoscopic group on G/Z_1 (in the sense of (2.1)) is that the character κ be trivial on Z_1.*

PROOF. By [17] κ restricted to Z_1 for quasi-split groups is independent of the Cartan subgroup T with endoscopic group H. The short exact sequence

$$1 \rightarrow Z_1 \rightarrow T \rightarrow T/Z_1 \rightarrow 1$$

gives

$$H^1(Z_1) \rightarrow H^1(T) \rightarrow H^1(T/Z_1).$$

By exactness (and the vanishing of an appropriate Ext^1) κ is trivial on $H^1(Z_1)$ if and only if it extends to a character κ_{T/Z_1} on $H^1(T/Z_1)$. If it extends then we may define an endoscopic group by the Cartan subgroup T/Z_1 and character κ_{T/Z_1}. If H descends to G/Z_1 it defines a character κ' on $H^1(T/Z_1)$ that restricts to κ on $H^1(T)$. By exactness κ restricted to $H^1(Z_1)$ is trivial.

REMARK 2.3. For a given endoscopic group H of a quasi-split group G we may use this idea to calculate the characters of (VI.2.3) in terms of the splitting field of H. We make this explicit for split groups.

Let G be a split reductive group. We may work with ${}^LG^0$ instead of ${}^LG$ since the L-group is a direct product of ${}^LG^0$ by the Weil group. Let K be the smallest extension through which ρ factors. Suppose we have groups $H, {}^LH^0, T, {}^LT^0$, simply connected cover ${}^L\tilde{G}^0 \rightarrow {}^LG^0$ of ${}^LG^0$ with subgroups ${}^L\tilde{H}^0, {}^L\tilde{T}^0$ projecting to ${}^LH^0$ and ${}^LT^0$ respectively. Suppose also we have an element $s \in {}^LT^0$ with $Cent(s, {}^LG^0)^0 = {}^LH^0$ and $\tilde{s} \in {}^L\tilde{T}^0 \subseteq {}^L\tilde{H}^0$ projecting to s. The element $\rho(\sigma)$ can be written as $n(\sigma)|_{{}^LH^0}$ and lifted to $\tilde{n}_\sigma$ in $N_G({}^L\tilde{T}^0)$.

Since $n(\sigma)(s) = s$ we have $\tilde{n}_\sigma(\tilde{s}) = z_\sigma\tilde{s}$ where the element z_σ in $ker({}^L\tilde{G} \rightarrow {}^LG^0)$ depends only on $\rho(\sigma)$. This gives an injection $\sigma \rightarrow z_\sigma$ from $\mathrm{Gal}(K/F)$ to

$$ker({}^L\tilde{G} \rightarrow {}^LG^0).$$

By passing to a cover ${}^LG'^0$ of ${}^LG^0$ we may assume that G' is as "adjoint as possible", that is, the elements z_σ generate $ker({}^L\tilde{G}^0 \to {}^LG'^0)$ and that

$$\mathrm{Gal}(K/F) \to ker({}^L\tilde{G}^0 \to {}^LG'^0)$$

is an isomorphism. Identifying $ker({}^L\tilde{G}^0 \to {}^LG'^0)$ with

$$Hom(X^*({}^L\tilde{T}^0)/X^*({}^LT'^0), \mathbb{C}^\times)$$

we obtain an isomorphism

$$\rho : \mathrm{Gal}(K/F) \to Hom(X^*({}^L\tilde{T}^0)/X^*({}^LT'^0), \mathbb{C}^\times).$$

We identify $\mathrm{Gal}(K/F)$ with the dual of $X_*(T_{adj})/X_*(T')$. Now for a cyclic extension select a generator σ_1 of K. K will be a cyclic extension except possibly for $G = D_n$ where two generators σ_1, σ_2 might be needed. Then $\rho(\sigma_1)$ gives a character of $X_*(T_{adj})/X_*(T')$ which by Tate-Nakayama we identify with a character θ on $H^1(\mathrm{Gal}(K/F), Z')$ where Z' is the center of G'. To obtain the character on $H^1(\mathrm{Gal}(K/F), Z)$ we pull back the character on $H^1(\mathrm{Gal}(K/F), Z')$ by the map $Z \to Z'$. Note that the order of θ is precisely the order of σ in $\mathrm{Gal}(K/F)$. From construction it is clear that the character depends only on the endoscopic group and not the choice of Cartan subgroup.

Thus loosely speaking the characters on $H^1(Z)$ are determined by selecting the most adjoint possible group G' with a given endoscopic group, and selecting characters which generate the dual of $H^1(Z')$.

3. Compatibility of Characters.

As an application of the formulas for the subregular germ we check the compatibility of characters for the subregular germ. A necessary condition for the matching of the subregular germs is that the characters of the subregular germ of a κ-orbital integral match the characters of the subregular germ of a stable orbital integral on H.

Let G be a reductive group, T a Cartan subgroup, κ a character on T, and H an endoscopic group attached to the triple (G, T, κ). We have an asymptotic expansion along a regular curve Γ

$$\sum |\lambda|^{\beta-1}\theta(\lambda)F(\beta, \theta, f)$$

of the κ-orbital integral on T. We let $Y(T, \kappa)$ be the set of characters θ for which there exists an $f \in C_c(G)$ with $F(2, \theta, f) \neq 0$. Similarly, the stable orbital integral on H gives an expansion

$$\sum |\lambda|^{\beta-1}\theta(\lambda)F(\theta, \beta, f^H)$$

and we let $X(H)$ be the set of characters θ for which there exists an $f' \in C_c(H)$ with $F'(2, \theta, f') \neq 0$. $X(H)$ is independent of the Cartan subgroup T provided T is selected so that the subregular germs do not vanish (3.2). The sets $Y(T, \kappa)$ and $X(H)$ are independent of the regular curve Γ. The purpose of this section is prove (assuming a minor assumption about ${}^2E_6$).

THEOREM 3.1. *Let G be a quasi-split reductive group such that $G(F)$ contains a subregular unipotent element. If H is an endoscopic group attached to the pair (T,κ) then $Y(T,\kappa) = X(H)$ provided $X(H) \neq \phi$.*

The hypothesis that G is quasi-split is made to simplify the arguments. We begin with

LEMMA 3.2. *Suppose H is a quasi-split simple reductive group. Then $X(H) = \phi$ or $\{1\}$ provided $H \neq {}^2A_{2n}$. $X({}^2A_{2n}) = \{\eta_K\}$.*

PROOF. If $G \neq {}^2A_{2n}$ the characters $\kappa(E_\alpha)$ are trivial for all α and the result follows. If $G = {}^2A_{2n}$ we have the two term asymptotic expansion of the subregular germ given by $(V.4.1)$.

$$(1/2)|\lambda| \int |dX/X| \int h_2|\nu_2| + (1/2)|\lambda|\eta_K(\lambda) \int |dX/X| \int \eta_K(b_\sigma)h_2|\nu_2|.$$

We show that $\int h_2|\nu_2| = 0$. The form ν_2 of $(V.4.1)$ is

$$\omega x(\alpha)x(\beta)/(\lambda^2 dx(\alpha)dx(\beta))$$

whereas the form ω_E on E_α is given by

$$\omega_E = \omega x(\alpha)/(\lambda^2 dx(\alpha)).$$

Write $E_{\alpha,\beta,u} = E_\alpha(u) \cap E_\beta$. Thus we may obtain the form on ν_2 on $E_{\alpha,\beta,u}$ by taking the 2-form $\delta^{-1}d\xi dw/(\xi w^2)$ on $E_\alpha(u)$, dividing by $dx(\beta)/x(\beta)$ and restricting to $E_{\alpha,\beta,u}$. $x(\beta)d\xi/(\xi dx(\beta)) = 1$ on $E_\alpha(u) \cap E_\beta(u)$ so we may take our 1-form on $E_{\alpha,\beta,u}$ to be dw/w^2. w need not be a coordinate over F. But $v = w/(Rw+1)$ for some $R \in K_X$ will be a coordinate over F. Thus the principal value integral is by [16]

$$\int |dv/v^2| = 0.$$

This proves that $X({}^2A_{2n}) = \{\eta_K\}$.

PROOF OF 3.1. We work with the quasi-split groups. We use $(VI.2.4)$ together with the previous section to determine the characters θ. We rely heavily on the vanishing theorems $(VI.3.2)$ and $(VI.3.3)$ to eliminate unwanted terms. $(VI.3.2)$ and $(VI.3.3)$ tell us that if a component $E(\ell_\alpha, u)$ makes a contribution then one of the following conditions hold:

1) ℓ_α intersects three other lines in the Dynkin curve
2) ℓ_α intersects two lines that are interchanged by $\mathrm{Gal}(\bar{F}/F)$.
3) ℓ_α intersects a line ℓ_β and $\kappa^\alpha = \kappa^\beta$.

By $(VI.2.3)$, condition (3) implies that $\kappa^\alpha = \kappa^\beta = 1$. There is also the obvious condition that $E(\ell_\alpha, u)$ contain rational points. Using these criteria one can read off the irreducible components $E(\ell_\alpha, u)$ that contribute to the subregular germ.

${}^1A_n$. If $\theta \neq 1$ then $Y(T,\kappa) = \phi$. Also if $\theta \neq 1$ then by (1.1), $X(H) = \emptyset$. So if $\theta \neq 1$ $X(H) = Y(T,\kappa) = \emptyset$. If $\theta = 1$ then $Y(T,\kappa) = \{1\}$ and $X(H) = \{1\}$.

^2A_n. Assuming that $X(H) \neq \phi$, we may take G to be $U(n)$ and H to be $U(j)\times U(n-j)$ (1.1). If n is odd then $X(H) = \{1, \eta_K\}$. Up to isogeny the endoscopic group is an endoscopic group of the adjoint group $U(n)_{adj}$ so the characters in chart ($VI.2.3$) are all trivial. However by lemma ($V.4.1$), the subregular germ of the asymptotic expansion contains the characters $\{1, \eta_K\}$. If n is even and j is even then $X(H) = \{1\}$. Again the endoscopic group up to isogeny is an endoscopic group of $U(n)_{adj}$ so that the characters in chart ($VI.2.3$) are trivial. Thus $X(T,\kappa) = \{1\}$. Finally we consider the case that n is even, j and $n-j$ are odd. Then $X(H) = \{\eta_K\}$. This time, however, H is not an endoscopic group of $U(n)_{adj}$ (up to isogeny). The element $s \in GL(n,\mathbb{C})$ defining the endoscopic group has odd determinant so that it does not pass to $SL(n,\mathbb{C})$.

We show that $\theta_1 = 1$, and $\theta_2 = \eta_K$ in ($VI.2.3$). We have $\prod \kappa^\alpha(\alpha(t')) = 1$. Let $t' = z^{\beta^v}\sigma(z)^{\sigma(\beta^v)}$ where $z \in K, \sigma \in \mathrm{Gal}(K/F)$, and $< \alpha_i, \beta^v >= 1$ if $i = 1$ and 0 otherwise. (Such a cocharacter exists in $U(n)$). This gives $\theta_1(z) = 1$ for $z \in K^\times$ so that $\theta_1 = 1$. Now $\theta_2 N = 1$, so that θ_2 is either the trivial character or η_K. If θ_2 were the trivial character then H (up to isogeny) would be an endoscopic group of the adjoint group $U(n)_{adj}(2.2)$, so $\theta_2 = \eta_K$.

B_n. If $\theta \neq 1$ then by ($VI.2.3$) and section 2 $Y(T,\kappa) = \{\eta_K^n\}$ where K is the splitting field of H. $H = {}^2A_{n-1-2j} \times {}^2(B_j \times B_j)$. The factor ${}^2(B_j \times B_j)$ does not contain any F-rational subregular elements. Thus the character comes from ${}^2A_{n-1-2j}$ and $X(H) = \{\eta_K^{n-2j}\} = \{\eta_K^n\}$. So $X(H) = Y(T,\kappa)$. If $\theta = 1$ then $Y(T,\kappa) = X(H) = \{1\}$.

C_n. $X(H) = Y(T,\kappa) = \{1\}$.

$D_n, {}^2D_n$. n even. $Y(T,\kappa) = \{1\}$ and we never obtain an even rank unitary group.

$D_n, {}^2D_n$. n odd. $Y(T,\kappa) = \{\theta^2\}$. $\theta^2 = 1$ if and only if H descends to an endoscopic group of $SO(2n)$. If $\theta^2 = 1$ we see that ρ is trivial. ${}^LSO(2n)^0 = SO(2n)$ and the unitary piece drops out. So $X(H) = \{1\}$. If $\theta^2 \neq 1$ then H does not descend to $SO(2n)$ so inside $SO(2n)$, $\rho(\sigma)(s) = -s$. Thus the orthogonal factors will be interchanged and A_j will be unitary (of even rank). Thus $X(H) = \{\eta_K\}$, $(\eta_K = \theta^2)$.

Exceptional Groups $E_6, E_7, E_8, G_2, F_4, {}^2E_6, {}^3D_4, {}^6D_4$.

By examining ($VI.2.3$) we see that $Y(T,\kappa) = \{1\}$. To prove compatibility of characters we must show that these exceptional groups do not have an endoscopic group with an even rank unitary group as a factor. If $G = G_2, E_8, G_2$ or F_4 this is easy: the centers of these groups do not contain an involution ($|Z(E_8)| = |Z(G_2)| = |Z(F_4)| = 1$). The outer automorphism of order three in ${}^3D_4$ and ${}^6D_4$ cannot give rise to a unitary group. But we have seen that the endoscopic groups of D_4 and ${}^2D_4$ do not have an even rank unitary group as a factor. Thus ${}^3D_4$ and ${}^6D_4$ do not either. This leaves E_7 and ${}^2E_6$. Unfortunately, I know of no simple proof to show that their endoscopic groups do not contain any even rank unitary factors. We sketch a case by case proof in the following paragraphs.

E_7. We take the dual of E_7 to be the adjoint group of type E_7 over $\mathbb{C}$. By the result of Golubitsky and Rothschild (1.7) the centralizer of s in E_7 stabilizes one of the following subalgebras: $A_1 \oplus D_6, A_1^3 \oplus D_4, A_1^7, A_2 \oplus A_5, A_2^3 \oplus T^1, A_7, E_6 \oplus T^1, T^7$. We discuss each of these algebras in turn. $A_1 \oplus D_6$ gives no even rank unitary factors because D_6 does not. Similarly ${A_1}^3 \oplus D_4$ gives no even rank unitary factors. A_1^7 and T^7 can be immediately dismissed. $A_2 \oplus A_5$ requires some attention especially because the outer automorphism of $A_2 \oplus A_5$ which acts on both factors is realized in the group E_7. We must show that this outer automorphism does not lie in the centralizer of s where $Lie(Cent(s, E_7)) = A_2 \oplus A_5$. The extended diagram of E_7 is

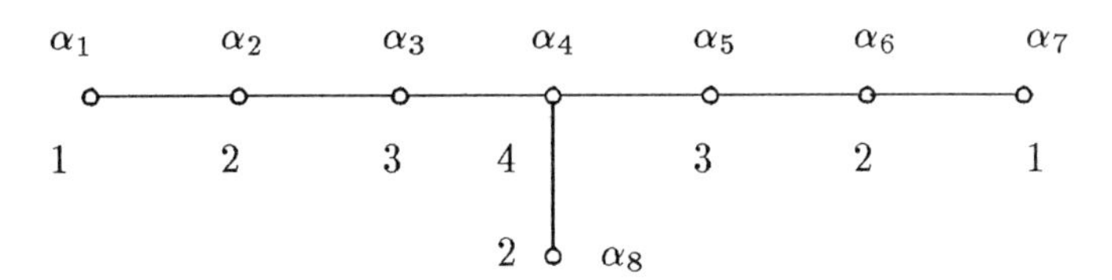

The element s is defined by $\alpha_i(s) = 1$ for $i \neq 5$, $\alpha_5(s) = x$. Since α_5 has the weight 3 in the extended diagram it follows that $x^3 = 1$. Thus the cube of every element in the centralizer of s lies in the connected component. It follows that the outer automorphism of order two of $A_2 \oplus A_5$ does not lies in the centralizer of s.

Next we exclude the case $A_2^3 \oplus T^1$. $\alpha_3(s) = y$, $\alpha_5(s) = x$, $(xy)^3 = 1$, $\alpha_i(s) = 1$, $i \neq 3, 5$. If there is to be a subregular unipotent element in $H(F)$ then ρ must stabilize one of the components A_2. But then the centralizer actually stabilizes $A_2 \oplus A_5$($i.e.$ we may take $y = 1$) and we reduce to the previous case.

Consider A_7. $\alpha_8(s) = x, x^2 = 1$ and $\alpha_i(s) = 1$ for $i \neq 8$. The center of the centralizer of s has two elements (namely 1 and s). So we may take it to be $SL(8, \mathbb{C})/\mu_4$ where μ_4 are the fourth roots of unity. We show that ${}^2A_{2k-1} \times {}^2A_{7-2k}$ is not an endoscopic group of the group with L group ${}^LG^0 = SL(8, \mathbb{C})/\mu_4$. Let $\hat{s} = diag(x^a, y^{8-a})$ $x \neq y$ where the exponents indicate the number of factors. $1 = det(\hat{s}) = x^a y^{8-a}, x \neq y$. Following section 1 we have $w(\sigma(\hat{s})) = diag((x^{-1})^a, (y^{-1})^{8-a}) = \lambda\hat{s}, \lambda \in \mu_4$. Thus $x^2 = y^2 = \lambda \in \mu_4$. It follows that $x = -y$ and $x^8 = y^8 = 1$. The determinant is then $(x/y)^a y^8 = (-1)^a = 1$, so that a must be even.

$E_6 \oplus T^1$ reduces to the case E_6 if E_7 does not contain the outer automorphism of E_6 or ${}^2E_6$ if it does.

${}^2E_6$. We turn our attention to the group ${}^2E_6$. I am forced to assume at this point that the list of primitive algebras for the connected group E_6 is the same as the list for the semidirect product of E_6 by $\{1, \omega\}$ with two components. Again we take the adjoint group over $\mathbb{C}$ for the L-group. The extended Dynkin diagram is given by:

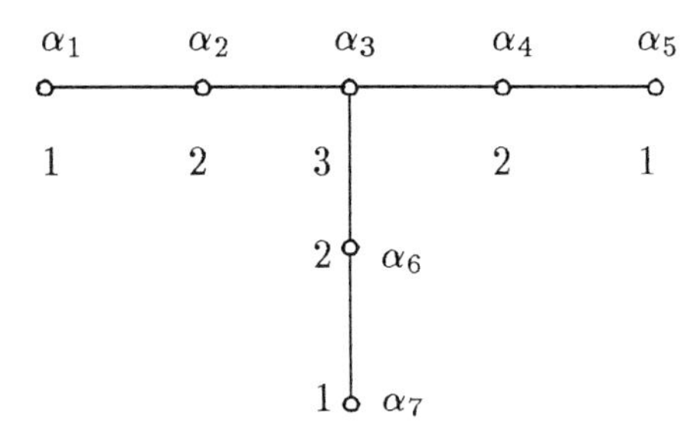

The centralizer stabilizes one of $A_1 \oplus A_5, A_2^3, D_4 \oplus T^2, D_5 \oplus T^1, T^6$. We begin with A_2^3. We identify the outer automorphisms of A_2^3 with signed permutations. Let x be the outer automorphism of E_6 considered as an automorphism of A_2^3, a signed permutation on three letters. Let y be an outer automorphism of A_2^3 of order 3 coming from $\mathrm{Cent}(s, {}^LG^0)$. With appropriate choices y and x may be represented by the signed permutations

$$y = \begin{pmatrix} 1 & 2 & 3 \\ 2 & 3 & 1 \end{pmatrix} \qquad x = \begin{pmatrix} 1 & 2 & 3 \\ \epsilon_1 2 & \epsilon_2 1 & \epsilon_3 3 \end{pmatrix}$$

x acts trivially on the factor of A_2^3 it stabilizes so $\epsilon_3 = 1$. y and x together generate a group isomorphic to the symmetric group on 3 letters so $xyx^{-1} = y^2$. An easy calculation shows this implies $\epsilon_1 = \epsilon_2 = 1$. Thus an outer automorphism that fixes a component acts as the trivial automorphism on that component.

$D_4 \oplus T^2$ is excluded because D_4 has no endoscopic groups with an even rank unitary factor. $D_5 \oplus T^1$ is a Levi component of E_6. Therefore the centralizer of s is simply connected inside the simply connected group of type E_6. But the orders of the centers of the simply connected groups of types E_6 and D_5 are relatively prime so that the centralizer of s is simply connected in the adjoint group as well. The group of type D_5 with a simply connected L-group does not have any endoscopic groups with unitary factors. T^6 is obviously excluded.

$A_1 \oplus A_5$ is the only remaining case. Let t be given by $\alpha_6(t) = -1$, $\alpha_i(t) = 1$ $i \neq 6$. The centralizer is seen to be

$$Cent(t, {}^LG^0)^0 = (SL(2, \mathbb{C}) \times SL(6, \mathbb{C})/\mu_3)/(\pm 1).$$

We work inside this subgroup. Proceeding as in the calculations for unitary groups we find that if we are to obtain a unitary factor over F then there must be a root α_j $(1 \leq j \leq 5)$ such that $\alpha_7(s) = -1$, $\alpha_i(s) = 1$ $(i \leq 5, i \neq j)$, $\alpha_j(s) = -1$. If the unitary factor is to have even rank j must equal $1, 3$, or 5. The weights in the extended diagram on these roots are odd so that $\alpha_j(s)^{m_j} = -1$. Using

$$\alpha_1(s)\alpha_2(s)^2\alpha_3(s)^3\alpha_4(s)^2\alpha_5(s)\alpha_6(s)^2\alpha_7(s) = 1$$

we find that $\alpha_6(s)^2 = 1$ so that $\alpha_6(s) = \pm 1$. Thus either $\alpha_6(s)$ or $\alpha_6\alpha_7(s) = 1$. Neither is a root of $Cent(t, {}^LG^0)^0$. This contradicts the assumption that ${}^LH^0$ stabilizes $A_1 \oplus A_5$.

4. Stable orbital integrals.

In this section we transfer the stable subregular germ on a group G to its quasi-split inner form G_{in}. We fix a measure by setting

$$\omega_{E_\alpha} x(\beta)/dx(\beta)|_{E_\alpha \cap E_\beta} = dw/w^2.$$

THEOREM 4.1. *For every subregular adjoint conjugacy class O in G there is a subregular adjoint conjugacy class O' in G_{in} such that $\Gamma_O = -\Gamma_{O'}$ where Γ_O and $\Gamma_{O'}$ are the germs of O and O' respectively.*

PROOF. Fix an adjoint conjugacy class O in G. There is only one component $E_\alpha(u)$ which contains any F-rational points $(VI.1.4)$. Select O' to be a subregular element in G_{in} such that the action of $\mathrm{Gal}(\bar{F}/F)$ on the lines of $(B\backslash G_{in})_{u'}$ $u' \in G(F)$ coincides with the action on the lines of $(B\backslash G)_u$ $u \in G(F)$. The only possible difference in the data for the two germs is the constant ζ that appears in formula $(VI.1.6)$. By blowing up as needed we obtain a morphism over F from $E_\alpha(u)$ to $\mathbb{P}^1$ given on $U(\alpha, \beta)$ in canonical coordinates by $(w, \xi) \rightarrow w$. Blowing up to extend the morphism does not affect principal value integrals because we never blow up at an F-rational point. The fibre over a given $p \in \mathbb{P}^1(F)$ does not necessarily have any rational points. It will have rational points if and only if the cocycle $\sigma(\xi)\xi^{-1}$ (depending on $p \in \mathbb{P}^1(F)$) in $H^1(U(1))$ is non-trivial. We introduce the non-trivial character η_K of $H^1(U(1))$. As p varies in $\mathbb{P}^1(F)$, $\sigma(\xi)\xi^{-1}$ equals a cocycle of $\mathrm{Gal}(\bar{F}/F)$ with coefficients in $U_{K_X(w)}(1)$. The integral over the fibre thus equals

$$\int |dX/X|.$$

The integral over the base equals

$$\int ((1 + \eta_K(\sigma(\xi)\xi^{-1})/2)|dw/w^2| = (1/2) \int \eta_K((\sigma(\xi)\xi^{-1})|dw/w^2|$$

because the principal value integral $\int |dw/w^2|$ is zero [16]. Now $\eta_K(\sigma(\xi)\xi^{-1}) = \eta_K(\zeta)\eta_K(*)$ where $*$ is independent of ζ and is consequently the same for G and G_{in}. It was proved in $(VI.1.6)$ that ζ is a norm if and only if G is quasi-split.

5. Unitary Groups.

In this section we prove the transfer of the subregular germ of κ-orbital integrals from unitary groups to the endoscopic groups $H = U(n - 2h) \times U(2h)$.

5.1. Vanishing of Germs.

We begin by proving that the subregular germ of a κ-orbital integral is zero if the endoscopic group H contains no F-rational points. There are two ways to prove this result. One is to use the formulas obtained in the last chapter. The principal value integrals can easily be shown to vanish. The other approach does not use Igusa theory, but calculates the action of $H^1(Z)$ on F-conjugacy classes in each adjoint conjugacy class. We take the second approach.

LEMMA 5.2. *Let G be a simply connected semi-simple group. The F-conjugacy classes in an adjoint class are in bijection with classes in the image of $H^1(F,Z)$ in $H^1(F,C_G(x)_{red})$ (where x is a fixed F-element of an adjoint class). In fact the classes are given by the image of $H^1(F,Z)$ in $H^1(F,(C_G(x)'_{red})$ where $C_G(x)'_{red}$ is the subgroup of $C_G(x)_{red}$ that acts trivially on components of $(B\backslash G)_x$.*

PROOF. Fix an F-conjugacy class O. Let a_σ be a cocycle in $H^1(F,Z)$. By Kneser [13], there is an element $g \in G(\bar{F})$ such that $\sigma(g)g^{-1}$ is a cocycle representing a_σ. Let O' be the conjugacy class O^g. It is adjointly conjugate to O, and it depends only on the class in $H^1(F,Z)$ and not on the choice of cocycle or element g. Suppose that O and O' are adjointly conjugate F-conjugacy classes. Then $O^g = O'$ for some g such that $\sigma(g)g^{-1} = z_\sigma \in Z^1(F,Z)$. Thus we have a map from $H^1(F,Z)$ onto the set of classes adjointly conjugate to O.

Suppose that a_σ acts trivially on O. This means there is a g such that $\sigma(g)g^{-1}$ represents a_σ and $O^g = O$. Adjusting by an element in $G(F)$, we may also assume that $x^g = x$ for some $x \in O$. Thus $\sigma(g)g^{-1}$ is a boundary in $C_G(x)$, and g lies in $C_G(x)$. $C_G(x)$ is a semi-direct product of a reductive piece $C_G(x)_{red}$ and its unipotent radical [21,7.5]. Write $g = g_0 n$ according to this semi-direct product.

$$\sigma(g)g^{-1} = \sigma(g_0)g_0^{-1} \cdot \mathbf{ad}(g_0)(\sigma(n)n^{-1}).$$

Since a_σ is central the cocycle $\sigma(g)g^{-1}$ lies in $C_G(x)_{red}$. This forces $\sigma(n)n^{-1}$ to be the identity so that we might as well take $\sigma(g_0)g_0^{-1}$ to be our representative of a_σ. Thus the kernel of the action of $H^1(F,Z)$ on adjoint conjugacy classes lies in the kernel of the map from $H^1(F,Z)$ to $H^1(F,C_G(x)_{red})$. It is easy to see that this is actually a bijection. For the last statement of the lemma, it suffices to remark that Z acts trivially on components so that the image actually lies in this smaller subgroup.

LEMMA 5.3. *Let $x \in G(F)$ be subregular, $G = SU(n)$, $(n \geq 3)$. κ is trivial on $ker(H^1(Z) \to H^1(C_G(x)_{red})$ if and only if H (up to isogeny) is an endoscopic group of $U(n)$.*

PROOF. By Slodowy [21], $C_G(x)_{red}$ is connected and is in fact isomorphic to $\mathbb{G}_m$ over $\bar{F}$. Let B_+ be a Borel subgroup over F in $(B\backslash G)_x$ (n odd) or let B_+ be one of the two Borel subgroups in ℓ_α lying at the intersection with a second line where ℓ_α is the line in $(B\backslash G)_x$ over F (n even). Since elements of $C_G(x)_{red}$ stabilize $(B\backslash G)_x$, $C_G(x)_{red} \subseteq B_+$. By (5.2), we see that the image of $H^1(Z)$ lies in $H^1(C_G(x)_{red})$. If n is odd fix a Cartan subgroup T over F in B_+ containing $C_G(x)_{red}$. If n is even we again fix a Cartan subgroup T in B_+ containing $C_G(x)_{red}$. It will no longer be defined over F. Let N be the unipotent radical of B_+. It is clear that if $b \in C_G(x)_{red}$ then

$$\alpha'(b \text{ modulo } N) = 1$$

if x does not lie in a line of type α'. If B_+ lies at the intersection of α_1 and α_2 then $x(\alpha_1 + \alpha_2) \neq 0$ so that

$$\alpha_1\alpha_2(b \text{ modulo } N) = 1.$$

Thus identifying T with the diagonal matrices and B_+ with the upper triangular matrices we see that b modulo N has the form

$$diag(a, a, \ldots, a, a^{-n+1}, a, \ldots, a).$$

The morphism $T \rightarrow U(1)$ defined by

$$diag(t_1, \ldots, t_n) \rightarrow t_1$$

yields an isomorphism over F of a subtorus of T with $U(1)$ even when the torus T is not defined over F. This gives an isomorphism over F of $C_G(x)_{red}$ with $U(1)$. If we identify $Z(\bar{F})$ with the nth roots of unity, then the morphism $Z \rightarrow C_G(x)$ corresponds to the inclusion $\mu_n \rightarrow U(1)$. The center of $U(n)$ (or any of its inner forms) is isomorphic over F to $U(1)$ and the inclusion $SU(n) \rightarrow U(n)$ gives an inclusion $Z \rightarrow U(1)$ or $\mu_n \rightarrow U(1)$. This demonstrates that the kernel of the homomorphism

$$H^1(Z) \rightarrow H^1(C_G(x)_{red})$$

equals the kernel of the homomorphism

$$H^1(Z) \rightarrow H^1(Z_{U(n)}).$$

From the formalities of centers, characters and endoscopic groups given in section 2, we know that H (up to isogeny) is an endoscopic group of $U(n)$ if and only if κ is trivial on this kernel. Hence the result.

LEMMA 5.4. *If κ is non-trivial on the $ker(H^1(Z) \rightarrow H^1(C_G(u)_{red})$ then the germ of the conjugacy class of u equals zero.*

PROOF. Define an action of $Z \backslash G(F)$ on $f \in C_c(G)$ by $z \cdot f(g) = f(z^{-1} g z)$. Then it is easy to see that with the proper normalization of measures

$$\kappa(\sigma(z^{-1})z)\Phi(f) = \Phi(z \cdot f) = \sum \Gamma_O \mu_O(z \cdot f) = \sum \Gamma_O \mu_{(O^z)}(f)$$

where $\Phi(f)$ is the κ-orbital integral of f, Γ_O is the germ of the unipotent conjugacy class O and μ_O is an invariant measure on O. By the uniqueness of germ expansions $\kappa(\sigma(z^{-1})z)\Gamma_O = \Gamma_{\mathbf{ad}z O}$. If κ is non-trivial on

$$ker(H^1(Z) \rightarrow H^1(C_G(u)_{red})$$

pick z such that $\mathbf{ad}zO = O$ and $\kappa(\sigma(z^{-1})z) \neq 1$. Then $\Gamma_O = 0$.

REMARK. This centralizer argument can be applied quite generally to show that germs vanish. One can show for instance for $G = A_n$ that if the splitting field of H is cyclic of order ℓ, then the only non-vanishing germs correspond to unipotent classes such that ℓ divides the lengths of all the Jordan blocks of an element of the conjugacy class. This implies, in particular, that the asymptotic expansion has the form

$$\sum |\lambda|^{\ell\beta} \theta(\lambda) F(\ell\beta, \theta, f).$$

where β is a non-negative integer.

5.5. The stable germ of unitary groups.

The next two sections contain calculations that give the transfer of the subregular germ of $G = U(n)$ to $H = U(n-2h) \times U(2h)$. The general idea should not be obscured by the calculations that follow. As was mentioned in chapter V, our expression of the subregular germ consists of the following data:
1. A surface S (together with a description of its components and rationality structure)
2. a 2-form on S defined over F.
3. a cocycle b_σ in T depending on $p \in S(F)$
4. a character κ on T
5. canonical coordinates (w, ξ).

As in the transfer of stable germs, we will be able to integrate out the dependence on ξ at the expense of introducing a cocycle a_σ in $H^1(U(1))$ depending on w. Furthermore the cocycle in T will be simplified to a cocycle in $H^1(U(1))$. The data will then become:
1. a projective line $\mathbb{P}^1$ with canonical coordinate w
2. a 1-form dw/w^2 on $\mathbb{P}^1$
3. cocycles b and $a \cdot b$ in $H^1(U(1))$
4. the non-trivial character η_K of $H^1(U(1))$

The germ is related to the data on $U(n)$ by the formula

$$(1/2)|\lambda| \int \eta_K(b)|dw/w^2| + (1/2)|\lambda|\eta_K^n(\lambda) \int \eta_K(a \cdot b)|dw/w^2|.$$

Here η_K^n is the nth power of the character η_K. The cocycle $\mathbf{a}$ depends on the rank of the group so we add a subscript n when discussing more than one unitary group. The cocycle $\mathbf{b}$ will depend on G and on $H = U(n-2h) \times U(2h)$. We add subscripts to indicate this dependence $\mathbf{b} = b_{n,2h}(2h \leq n)$. In particular $b_{n,0}$ corresponds to the stable orbital integral (H $= U(n)$). We also add subscripts to the variables (w, ξ) and to the form ν. The transfer will follow from the following three steps:
1. If κ is trivial (*i.e.* for stable orbital integrals) $b_{n,0} = 1$ and

$$\int \eta_K(b_{n,o})|dw/w^2| = 0.$$

2. (Transfer to $U(2h)$) There is a morphism $\mathbb{P}^1 \to \mathbb{P}^1$ over F given by

$$w_n/(R_1 w_n + 1) = w_{2h} \text{ for some } R_1 \in K_X$$

carrying $b_{n,2h}$ to a_{2h}. (Note that $\eta_K^{2h} = 1$ and that $dw_n/w_n^2 = dw_{2h}/w_{2h}^2$).
3. (Transfer to $U(n-2h)$) There is a morphism $\mathbb{P}^1 \to \mathbb{P}^1$ over F given by

$$w_n/(R_2 w_n + 1) = w_{n-2h} \text{ for some } R_2 \in K_X$$

carrying $a_n b_{n,2h}$ to a_{n-2h}. (Note that $\eta_K^n = \eta_K^{n-2h}$).

We must also discuss the degenerate case $n-2h=1$ when n is odd.

Let $\epsilon=1$ when n is even and $\epsilon=0$ when n is odd. We write element of the Cartan subalgebra as $(x_k,\dots,x_{-k})$ $(2k+1-\epsilon=\mathrm{n})$ with the understanding that $x_0=0$ if n is even. We let T_i, $-k\le i\le k$ to be the character on the Cartan subalgebra given by $T_i\ (x_k,\dots,x_{-k})=x_i$. The Weyl group may be identified with permutations on $2k+1-\epsilon$ letters $T_k,\dots,T_{-k}$. Positive roots are identified with T_i-T_j $i>j$ $(i,j\neq 0$ if n is even). We write ϵ_j $j\geq 1$ for the permutation $(j,-j)$. The field K_X introduced in $(V.2)$ is $\bar{F}(T_k,\dots,T_{-k})$.

We give a description of the stable subregular germ of $U(n)$.

THEOREM 5.6. *Let α and β be the simple roots $\alpha=T_1-T_{-\epsilon},\beta=T_{-\epsilon}-T_{-\epsilon-1}$. The stable subregular germ of $U(n)$ is given by the following data*
1. *The curve $\mathbb{P}^1$ with canonical variable w. The action of $\tilde{\Omega}$ on w is given by*

$$\begin{aligned}\sigma_{\alpha'}(w)&=w \qquad \alpha'\neq\alpha,\beta\\ \sigma_{\alpha}(w)&=w/(\alpha(X)e(\alpha,\beta)w+1)\\ \sigma_{\beta}(w)&=w/(\beta(X)e(\beta,\alpha)w+1)\\ \sigma_{0}(w)&=-w/((T_\epsilon-T_{-\epsilon})e(\alpha,\beta)w+1)\end{aligned}$$

2. *The 1-form dw/w^2*
3. *The cocycle a of $\tilde{\Omega}$ with values in $U(1)$ given on generators by*

$$\begin{aligned}&\sigma_{\alpha'}\to 1 \quad \textit{if} \quad \alpha'\neq\alpha\\ &\sigma_{\alpha}\to(\alpha(X)e(\alpha,\beta)w+1)\\ &\sigma_0\to 1/\zeta, \qquad \zeta\in F^\times : n \textit{ even}\\ &\sigma_0\to w/x(\gamma) : n \textit{ odd.}\end{aligned}$$

REMARK. We take $e(\alpha,\beta)=1$, $e(\alpha,\beta')=-1$, $e(\beta,\alpha)=-1$, $(\beta'=T_2-T_1)$. This is justified by the 3×3 calculations:

$$\begin{pmatrix}1&0&0\\1&1&0\\0&0&1\end{pmatrix}\begin{pmatrix}1&0&1\\0&1&0\\0&0&1\end{pmatrix}\begin{pmatrix}1\,0\,0\\-1\,1\,0\\0\,0\,1\end{pmatrix}=\begin{pmatrix}1&0&1\\0&1&\underline{1}\\0&0&1\end{pmatrix}\leftarrow e(\alpha,\beta)$$

and similarly for $e(\beta,\alpha)$ and $e(\alpha,\beta')$.

PROOF. Begin with n odd. By $(V.4.1)$ and section 4 the subregular germ is given by

$$(1/2)|\lambda|^2\eta_K(\lambda)\int|dX/X|\int\eta_K(a_\sigma)|dw/w^2|.$$

The action of $\tilde{\Omega}$ on the variables is given in $(VI.1.6)$ and $(VI.1.7)$. The cocycle a_σ for n odd is given by $(V.4)$

$$\begin{aligned}&\sigma\to\lambda/(\sigma(x(\beta))x(\beta)) &&(\sigma \text{ not in } \Omega\subseteq\tilde{\Omega})\\ &\sigma\to\sigma(x(\beta))x(\beta)^{-1} &&(\sigma \text{ in } \Omega).\end{aligned}$$

For simple roots $\sigma_{\alpha'}(x(\beta))/x(\beta) = x(W(\sigma_{\alpha'}), \beta)/x(\beta)$ and we apply $(V.6.1)$. For $\sigma \to \sigma_0$, $\sigma_0(x(\beta)) = x(\alpha)$,

$$\lambda/\sigma(x(\beta))x(\beta) = x(\alpha)x(\beta)w/(x(\alpha)x(\beta)x(\gamma)) = w/x(\gamma).$$

When n is even we integrate out the contribution of ξ. By blowing up if necessary we extend the morphism $(w, \xi) \to (w)$ to a morphism $E_\alpha(u) \to \mathbb{P}^1$. We integrate over exactly those $w \in \mathbb{P}^1$ such that the fibre over w has rational points. Now $a_\sigma = \sigma([\xi])[\xi]^{-1}$ is a cocycle of $\tilde{\Omega}$ in $U(1)$ which is trivial exactly when the fibre has rational points. When the fibre has rational points the integral over the fibre is $\int |dX/X|$. Thus the germ may be written in the form

$$\int |dX/X| \int (1/2)(1 + \eta_K(a_\sigma))|dw/w^2|.$$

That a_σ has the form given in the lemma follows immediately from lemma $(VI.1.6)$.

To carry out a comparison of orbital integrals we must switch to a different set of generators of $\tilde{\Omega}$. We let $\sigma_{\alpha'} = (\ell+1, \ell)$ act on w by

$$(\ell+1, \ell)(w) = w/((T_{\ell+1} - T_\ell)e_\ell w + 1).$$

Here e_ℓ are for now arbitrary constants in F.

LEMMA 5.7. *This action extends uniquely to an action of Ω on $K_X(w)^\times$.*

PROOF. Since simple reflections generate Ω, the extension if it exists is necessarily unique. We must check that if $\sigma(w) = w/(A_\sigma w + 1)$ then

$$w/(A_{\sigma\tau} w + 1) = \sigma\tau(w) = \sigma(w/(A_\tau w + 1) = w/((\sigma(A_\tau) + A_\sigma)w + 1).$$

That is, we must check that A_σ is a cocycle. Since Ω is a Coxeter group it is sufficient to verify that $A_{(\sigma_\alpha\sigma_\beta)^3} = 1$ for α and β adjacent simple roots and $A_{(\sigma_\alpha^2)} = 1$ for α simple.

$$\begin{aligned}
A_{(\ell+1)^2} &= (\ell+1, \ell)A_{(\ell+1,\ell)} + A_{(\ell+1,\ell)} \\
&= (\ell+1, \ell)(T_{\ell+1} - T_\ell)e_\ell + (T_{\ell+1} - T_\ell)e_\ell = 0 \\
A_{(\ell+2,\ell)} &= A_{(\ell+1,\ell)(\ell+2,\ell+1)(\ell+1,\ell)} \\
&= (\ell+1, \ell)(\ell+2, \ell+1)A_{(\ell+1,\ell)} + (\ell+1, \ell)A_{(\ell+2,\ell+1)} + A_{(\ell+1,\ell)} \\
&= (\ell+1, \ell)(\ell+2, \ell+1)(T_{\ell+1} - T_\ell)e_\ell \\
&\qquad + (\ell+1, \ell)(T_{\ell+2} - T_{\ell+1})e_{\ell+1} + (T_{\ell+1} - T_\ell)e_\ell \\
&= (T_{\ell+2} - T_{\ell+1})e_\ell + (T_{\ell+2} - T_\ell)e_{\ell+1} + (T_{\ell+1} - T_\ell)e_\ell \\
&= (T_{\ell+2} - T_\ell)(e_\ell + e_{\ell+1}) \\
A_{(\ell+2,\ell)^2} &= (\ell+2, \ell)(T_{\ell+2} - T_\ell)(e_\ell + e_{\ell+1}) + (T_{\ell+2} - T_\ell)(e_\ell + e_{\ell+1}) = 0.
\end{aligned}$$

LEMMA 5.8. $A_{(\ell,-\ell)} = (T_\ell - T_{-\ell})(e_{\ell-1} + \ldots + e_{-\ell}) \quad (\ell \geq 1)$.

PROOF. When $\ell = 1$ we obtain the result by the calculation of $A_{(\ell+2,\ell)}$ carried out in the proof of the previous lemma. Let r be the permutation $r = (j+1,j)(-j,-1-j)$ $j \geq 1$. Then $r(j,-j)r = (j+1,-j-1)$. So that

$$A_{(j+1,-j-1)} = r(j,-j)A_r + rA_{(j,-j)} + A_r$$
$$A_r = A_{(j+1,j)} + A_{(-j,-j-1)} = (T_{j+1} - T_j)e_j + (T_{-j} - T_{-j-1})e_{-j-1}$$
$$\text{So } r(j,-j)A_r + A_r = (T_{j+1} - T_{-j-1})(e_j + e_{-j-1}).$$

By induction we may assume that $A_{(j,-j)} = (T_j - T_{-j})(e_{j-1} + \ldots + e_{-j})$. We conclude that $A_{(j+1,-j-1)} = (T_{j+1} - T_{-j-1})(e_j + \ldots + e_{-j-1})$.

To apply this result to $U(n)$ when n is odd we set $e_i = 0$ $i \neq 0,-1$, $e_0 = e(\alpha,\beta)(= 1)$, $e_{-1} = e(\beta,\alpha)(= -1)$. Then this action corresponds to the action on w given in (5.6). Thus $e_0 + e_{-1} = 0$ so that $A_{(\ell,-\ell)} = 0$ for all $\ell \geq 1$. When n is even, we let $(e_0 + e_{-1}) = e(\alpha,\beta), e_{-2} = e(\beta,\alpha)$ and $e_i = 0$ otherwise. Then $A_{(\ell,-\ell)} = 0$ for $\ell \geq 2$ and $A_{(1,-1)} = (T_1 - T_{-1})e(\alpha,\beta)$. This proves:

COROLLARY 5.9. *$A_{(\ell,-\ell)} = 0$ for all $\ell \geq 2$. $A_{(1,-1)} = (T_\epsilon - T_{-\epsilon})e(\alpha,\beta)$.*

We suppress the direction X from the notation in most of what follows.

LEMMA 5.10. *$a_{\epsilon_j} = ((T_j - T_{-\epsilon})we(\alpha,\beta) + 1)/((T_{-j} - T_{-\epsilon})we(\alpha,\beta) + 1)$ if $j \geq 1 + \epsilon$ and $a_{\epsilon_1} = (T_1 - T_{-1})we(\alpha,\beta) + 1$ if $j = \epsilon = 1$.*

PROOF. Suppose that n is odd then $\epsilon_1 = \sigma_\beta\sigma_\alpha\sigma_\beta$. By (5.6) $a_{\sigma_\alpha} = (T_1 - T_0)e(a,\beta)w + 1$ and $a_{\sigma_\beta} = 1$. Thus

$$a_{\epsilon_1} = \sigma_\beta a_{\sigma_\alpha} = \sigma_\beta((T_1 - T_0)e(\alpha,\beta)w + 1) = (T_1 - T_{-1})e(\alpha,\beta)\sigma_\beta(w) + 1 =$$
$$((T_1 - T_0)e(\alpha,\beta)w + 1)/((T_{-1} - T_0)e(\alpha,\beta)w + 1).$$

Suppose that n is even then $\epsilon_1 = \sigma_\alpha$ so that (5.6)

$$a_{\epsilon_1} = (T_1 - T_{-1})e(\alpha,\beta)w + 1.$$

Let $r = (1,2)(-1,-2)$. Then $a_r = 1$ and $\epsilon_2 = r\epsilon_1 r$ so that $a_{\epsilon_2} = r(a_{\epsilon_1})$.

$$r(a_{\epsilon_1}) = (1,2)(-1,-2)((T_1 - T_{-1})e(\alpha,\beta)w + 1) =$$
$$(T_2 - T_{-2})e(\alpha,\beta)\sigma_\beta(w) + 1 =$$
$$((T_2 - T_{-1})e(\alpha,\beta)w + 1)/((T_{-2} - T_{-1})e(\alpha,\beta)w + 1).$$

Suppose that $a_{\epsilon_j} = ((T_j - T_{-\epsilon})e(\alpha,\beta)w + 1)/(T_{-j} - T_{-\epsilon})e(\alpha,\beta)w + 1)$ $j \geq 1 + \epsilon$ then

$$a_{\epsilon_{j+1}} = (j,j+1)(-j,-j-1)a_{\epsilon_j} =$$
$$((T_{j+1} - T_{-\epsilon})e(\alpha,\beta)w + 1)/((T_{-j-1} - T_{-\epsilon})e(\alpha,\beta)w + 1)$$

and the result follows by induction.

5.11. The κ-subregular germ on $U(2n)$.

In this section we calculate a simple expression for $m_\kappa(e)$ when $G = U(n)$ or an inner form thereof. We assume in the remainder of this chapter that $G(F)$ contains a subregular unipotent element and $H = U(2h) \times U(n-2h) = H_1 \times H_2$. We let $\underline{H}$ be the subgroup $U(2h) \times U(n-2h)$ of G with the following form

$$\begin{pmatrix} h & 0 & h \\ 0 & n-2h & 0 \\ h & 0 & h \end{pmatrix}.$$

If n is odd G is quasi-split if $G(F)$ contains a subregular unipotent element so these subgroups exist. If $G(F)$ contains a subregular unipotent element we may assume that the cocycle defining the inner form lies in a parabolic subgroup of type α, in fact we may assume it lies inside a Levi component M of such a parabolic subgroup. Thus again the subgroups $\underline{H}$ exist over F. If H' is a subgroup of G over F which is stably conjugate to $\underline{H}$ then $\underline{H}^g = H'$ for some $g \in G_{der}(\bar{F})$. We find that $\sigma(g)g^{-1}$ is a cocycle in $\underline{H}$ which lies in G_{der}.

LEMMA 5.12. *Let det_i be the determinant on $\underline{H}_i$. Then H' is stably conjugate to $\underline{H}$ if and only if $det_1(\sigma(g)g^{-1})$ is the trivial cocycle in $H^1(U(1))$.*

PROOF. We have the short exact sequence

$$1 \to SU(2h) \times SU(n-2h) \to U(2h) \times U(n-2h) \xrightarrow{(det_1, det_2)} U(1)_1 \times U(1)_2 \to 1.$$

This gives an injection

$$H^1(U(2h) \times U(n-2h)) \to H^1(U(1) \times U(1))$$

because by Kneser [13] $H^1(SU(2h) \times SU(n-2h)) = 1$. Moreover, this injection is actually an isomorphism, for $U(n) \to U(1)$ has a section. If $n = 2k+1$ the section is $x \to diag(1^k, x, 1^k)$. If n is even $U(2) \subseteq U(n)$ and $U(2)$ contains a torus isomorphic to $U(1)$. Since $\sigma(g)g^{-1}$ lies in G_{der}, its image lies in the diagonal $(x, x^{-1}) \in U(1) \subseteq U(1)_1 \times U(1)_2$. Thus $\sigma(g)g^{-1}$ gives a cocycle in $H^1(U(1))$. H' is conjugate to H over F if and only if $\sigma(g)g^{-1}$ gives the trivial class of $H^1(U(1))$. We conclude that the subgroups stably conjugate to $\underline{H}$ modulo F-conjugacy are in bijection with elements of $H^1(U(1))$.

LEMMA 5.13.

a) *The Cartan subgroups in G which are identified with Cartan subgroups in H are precisely those which are stably conjugate to a Cartan subgroup in $\underline{H}$.*

b) *Selecting a representative T in $\underline{H}$ for each of these stable conjugacy classes, if $g \in T\backslash G(F)$, then $\kappa(\sigma(g)g^{-1}) = \eta_K(det_1(\sigma(g)g^{-1}))$ where η_K is the non-trivial character on $H^1(U(1))$.*

PROOF. (a) is sufficiently clear.

(b) Suppose that $h \in (T\backslash H)(F)$. We show that $\kappa(\sigma(h)h^{-1}) = 1$. Let T_d be the subgroup $T \cap \underline{H}_{der}$ of T. Then we may take $\sigma(h)h^{-1}$ to lie in $H^1(T_d)$. By Tate-Nakayama we have the commutative diagram

$$\begin{array}{ccccc} H^1(T_d) & \xrightarrow{\sim} & H^{-1}(X_*(T_d)) & \xrightarrow{\sim} & H^{-1}(X^*({}^LT_d^0)) \\ \downarrow & & \downarrow & & \downarrow \\ H^1(T) & \xrightarrow{\sim} & H^{-1}(X_*(T)) & \xrightarrow{\sim} & H^{-1}(X^*({}^LT^0)). \end{array}$$

We have $s \in {}^L T^0$ corresponding to κ on $H^1(T)$. Since $\underline{H}_{der}$ is simply connected ${}^L(\underline{H}_{der})^0$ is adjoint. We have dual morphisms

$$\begin{array}{ccc} {}^L\underline{H}^0 & \rightarrow & {}^L(\underline{H}_{der})^0 \\ \uparrow & & \uparrow \\ {}^L T_0 & \rightarrow & {}^L(T_d)^0 \end{array}$$

$s \in {}^L T_0$ is central in ${}^L\underline{H}^0$ so that the image of s in ${}^L(T_d)^0$ is central in ${}^L(\underline{H}_{der})^0$. But ${}^L(\underline{H}_{der})^0$ is adjoint so the image of s in ${}^L(T_d)^0$ is the identity. It is now clear that $\kappa(\sigma(h)h^{-1}) = 1$. (b) follows immediately.

The next step is to compute the determinant of the cocycle $\sigma(g)g^{-1}$. To completely determine the cocycle it is enough to calculate it for generators. As in (5.5) we have characters T_i, $-k \leq i \leq k$ $(2k+1-\epsilon = n$. The Weyl group of $\underline{H}$ is then generated by the simple reflections $(i, i+1)$ $i \neq k-h, -k+h-1$ together with the involution $\epsilon_{k-h+1} = (k-h+1, -k+h-1)$. Thus it is sufficient to calculate the determinant on these generators together with the outer automorphism σ_0. We let γ_j be the positive root $T_j - T_{-j}$, $j = 1, \ldots, k$.

LEMMA 5.14. *On the regular elements $Y^0(B_0, B_\infty)$ up to a factor in $\bar{F}^\times$ independent of the star the cocycle $det_1(\sigma(g)g^{-1})$ is given by*

$$b_\sigma = 1 : \ \sigma \rightarrow (i, i+1) \quad i \neq k-h, -k+h-1$$

$$n_{\gamma_{n-h+1}}/m_{\gamma_{n-h+1}} : \sigma \rightarrow \epsilon_{n-h+1}$$

where $n = \prod \epsilon_\gamma(n_\gamma)$ and $n^{-1} = m = \prod \epsilon_\gamma(m_\gamma)$. The order on the roots is that given in (II.5).

PROOF. By lemma (I.5.4), $\sigma(g)g^{-1} = z(W_+, \alpha)^{\alpha^\vee}$ for $\sigma \rightarrow \sigma_\alpha = (i, i+1)$ (up to a constant independent of the star). If $i \neq k-h, -k+h-1$ then $det_1(b_\sigma) = det_1 z(W_+, \alpha)^{\alpha^\vee} = 1$. Next we compute b_σ for $\sigma \rightarrow \omega = \epsilon_1\epsilon_2 \ldots \epsilon_{n-h+1}$. By (I.5.6) it is sufficient to calculate the principal minors of $n^{-1}\omega \in N_\infty N_0 T_\sigma^\omega$. To recover $det_2(\sigma(g)g^{-1}) = det_1(\sigma(g)g^{-1})^{-1}$ it is sufficient to compute the $n-h^{th}$ and h^{th} principal minors. $n^{-1}\omega$ has the form

$$\begin{pmatrix} 1 & m_{ij} & m_{ij} \\ 0 & 1 & m_{ij} \\ 0 & 0 & 1 \end{pmatrix} \begin{pmatrix} 1_{h-1} & 0 & 0 \\ 0 & J_{n-2h+2} & 0 \\ 0 & 0 & 1_{h-1} \end{pmatrix}.$$

We see that the hth principal minor is

$$\begin{vmatrix} 1 & * & * \\ 0 & 1 & * \\ 0 & 0 & m_\gamma \end{vmatrix} = m_\gamma.$$

The $n-h^{th}$ principal minor equals the $n-h^{th}$ principal minor in the following $n-h+1$ by $n-h+1$ matrix

$$\begin{vmatrix} m_{ij} & & 1 \\ & 1 & 0 \\ 1 & 0 & 0 \end{vmatrix}$$

which is plus or minus the cofactor of m_γ in

$$\begin{vmatrix} 1 & & m_{ij} & \\ 0 & 1 & & \\ 0 & 0 & 1 & \end{vmatrix}.$$

So the $n-h^{th}$ principal minor equals $\pm n_\gamma$ $(\gamma=\gamma_{n-h+1})$.

The following lemma gives the restriction of m_γ/n_γ to the variety $E = E_{\alpha_k}\cap E_{\alpha_{k+1}}$ $({}^2A_{2k+1})$ or $E=E_{\alpha_k}$ $({}^2A_{2k})$. We work on the coordinate patch $U(\alpha_k,\alpha_{k+1})$ and use canonical coordinates.

LEMMA 5.15. *Up to a constant in $K_X^\times$ independent of (w,ξ), $n_{\gamma_p}/m_{\gamma_p}$ restricted to E equals*

$$((T_p-T_{-\epsilon})e(\alpha,\beta)w+1)/((T_{-p}-T_{-\epsilon})e(\alpha,\beta)w+1)$$

provided $p\neq 1$ if $\epsilon=1$. If $p=\epsilon=1, n_{\gamma_p}/m_{\gamma_p}$ is independent of (w,ξ).

REMARK 5.16. (Transfer factors for $H=U(n-2h)\times U(2h)$). By the remarks in the proof of (3.1) on $U(n)$ we see that $m_\kappa(e)=\Delta_\Gamma$ on E_0 (i.e. $\kappa^\alpha=1\ \forall\ \alpha$). $m_\kappa(e)$ extends to $E_\alpha\cap E_\beta$ (n odd) and E_α (n even) and depends on w. $w=0$ defines the intersection of $E_\alpha\cap E_\beta$ (resp. E_α) with E_0 so that we write $m_\kappa(e)=\Delta_\Gamma\eta_K(b_\sigma(w))$ where $b_\sigma(w)$ is a cocycle in $H^1(U(1))$ and $b_\sigma(0)$ is trivial in $H^1(U(1))$. In other words, if we normalize our cocycles so that $w=0$ gives the trivial cocycle the transfer factor is precisely what it should be.

PROOF. Let $\tilde n_\beta=\prod z(\alpha)^{m(\alpha)}n_\beta$. Then $(II.5.1)$ gives

$$(*)\qquad \begin{aligned} \lambda w(\gamma) &= \sum(-1)^j\gamma^{-1}(t)(1-\beta_j(t))\tilde n_{\beta_1}\dots\tilde n_{\beta_j} \text{ or}\\ w(\gamma) &= \sum\beta_1^{-1}\tilde n_{\beta_1}(-1)w(\gamma-\beta_1)+((1-\gamma^{-1})/\lambda)\tilde n_\gamma. \end{aligned}$$

By the proof of $(V.1.1.b)$, $w(\gamma)=0$ on E if γ is not simple and $\gamma\neq\alpha+\beta,\alpha=T_1-T_{-\epsilon},\beta=T_{-\epsilon}-T_{-\epsilon-1}$. So if $\gamma=(T_p-T_q)$ $q\neq-1-\epsilon,p>q$ then on E (using $w(\alpha')=1,\alpha'$ simple) $(*)$ becomes

$$0=\tilde n_{T_p-T_{q+1}}(-1)+\gamma(X)\tilde n_{T_p-T_q}.$$

Iterating this result for $q>-1-\epsilon$ we find that $\tilde n_{T_p-T_q}$ $(q>-1-\epsilon)$ is constant on E since $\tilde n_{\alpha'}=z(\alpha')/z(W,\alpha')=1/z_1(W,\alpha')=1/T(W,\alpha')$ for $\alpha'\neq\alpha,\beta$.

If $q=-1-\epsilon$ and $p>1$ then

$$\begin{aligned} 0&=\tilde n_{T_p-T_1}(-1)w+\tilde n_{T_p-T_{-\epsilon}}(-1)+\gamma(X)\tilde n_{T_p-T_{-1-\epsilon}}\\ \text{and } 0&=\tilde n_{T_p-T_1}(-1)+(T_p-T_{-\epsilon})\tilde n_{T_p-T_{-\epsilon}}. \end{aligned}$$

Subtracting w times the second from the first:

$$[(T_p-T_{-\epsilon})w+1]\tilde n_{T_p-T_{-\epsilon}}=\gamma(X)\tilde n_{T_p-T_{-\epsilon-1}}.$$

We conclude that up to a factor $*$ independent of w

$$\tilde{n}_{T_p - T_{-1-\epsilon}} = *((T_p - T_{-\epsilon})w + 1) \text{ and so also}$$
$$\tilde{n}_{T_p - T_{-p}} = *'((T_p - T_{-\epsilon})w + 1).$$

Now suppose that $p = 1$ and n is odd. Then $T_1 - T_{-1} = \alpha + \beta$. So

$$w = \tilde{n}_\alpha(-1) + (T_1 - T_{-1})\tilde{n}_{\alpha+\beta}$$
$$\tilde{n}_\alpha = 1/T(W, \alpha) = 1/(T_1 - T_0).$$

We conclude that $\tilde{n}_{T_1 - T_{-1}} = *((T_1 - T_0)w + 1)$. If $p = 1$ and n is even then $\alpha = T_1 - T_{-1}$, and $\tilde{n}_\alpha = \tilde{n}_{T_1 - T_{-1}}$ is independent of w.

The calculation for $\tilde{m}_\gamma$ follows the same lines. $\lambda w(\gamma)$ equals the γth coefficient of $t^{-1}\tilde{n}^{-1}t\tilde{n} = t^{-1}\tilde{m}t\tilde{m}^{-1}$ where $\tilde{n} = \prod \epsilon_\gamma(\tilde{n}_\gamma)$ and $\tilde{n}^{-1} = \tilde{m}$. By the proof of $(II.5.1)$ the βth coefficient of $\tilde{m}^{-1}$ equals

$$\sum (-1)^j \tilde{m}_{\beta_1}\tilde{m}_{\beta_2}\ldots\tilde{m}_{\beta_j}$$

where $\beta_i = (T_{a_i} - T_{a_{i+1}})$ $a_i > a_{i+1}$ and $\beta = T_{a_1} - T_{a_{j+1}}$. The βth coefficient of $t^{-1}\tilde{m}t$ is $\beta^{-1}\tilde{m}_\beta$. It follows that the γth coefficient of $(t^{-1}\tilde{m}t)\tilde{m}^{-1}$ equals

$$\sum (-1)^j \beta_0^{-1}\tilde{m}_{\beta_0}\tilde{m}_{\beta_1}\ldots\tilde{m}_{\beta_j} + \sum (-1)^{j+1}\tilde{m}_{\beta_0}\tilde{m}_{\beta_1}\ldots\tilde{m}_{\beta_j}$$

where $\beta_i = (T_{a_i} - T_{a_{i+1}})$ $a_i > a_{i+1}$ and $\gamma = T_{a_0} - T_{a_{j+1}}$. So $\lambda w(\gamma) =$

$$\sum (-1)^j (1 - \beta_1^{-1})\tilde{m}_{\beta_1}\tilde{m}_{\beta_2}\ldots\tilde{m}_{\beta_j} =$$

$$[\sum_{j\geq 2} (-1)^{j-1}(1 - \beta_1^{-1})\tilde{m}_{\beta_1}\ldots\tilde{m}_{\beta_{j-1}}]\ \tilde{m}_{\beta_j}(-1) + (-1)(1 - \gamma^{-1})\tilde{m}_\gamma =$$

$$[\sum \lambda w(\gamma - \beta_j)\tilde{m}_{\beta_j}(-1)] + (-1)(1 - \gamma^{-1})\tilde{m}_\gamma.$$

$$\text{So } w(\gamma) = \sum w(\gamma - \beta_j)\tilde{m}_{\beta_j}(-1) + (-1)(1 - \gamma^{-1})\tilde{m}_\gamma/\lambda.$$

Again $w(\gamma) = 0$ on E if γ is not simple and $\gamma \neq \alpha + \beta$. If $\gamma = T_p - T_q$ $p - 1 > q, p \neq 1$ then on E

$$0 = (-1)\tilde{m}_{T_{p-1} - T_q} + (-1)\gamma(X)\tilde{m}_{T_p - T_q}.$$

Iterating this result for $p < 1$ we find that $\tilde{m}_{T_p - T_q}$ is constant for $p < 1$.

If $p = 1$ and $q < -1 - \epsilon$ then

$$0 = \tilde{m}_{T_{-\epsilon} - T_q}(-1) + \tilde{m}_{T_{-\epsilon-1} - T_q}(-1)w + (-1)(T_1 - T_q)\tilde{m}_{T_1 - T_q} \text{ and}$$
$$0 = \tilde{m}_{T_{-\epsilon-1} - T_q}(-1) + (-1)(T_{-\epsilon} - T_q)\tilde{m}_{T_{-\epsilon} - T_q}.$$

Multiplying the first equation by $T_{-\epsilon} - T_q$ and subtracting we obtain

$$0 = ((T_q - T_{-\epsilon})w + 1)\tilde{m}_{T_{-\epsilon-1} - T_q} + (-1)(T_1 - T_q)(T_{-\epsilon} - T_q)\tilde{m}_{T_1 - T_q}$$

so that up to a factor independent of w

$$\tilde{m}_{T_1-T_q} = *((T_q - T_{-\epsilon})w + 1)$$
$$\tilde{m}_{T_p-T_{-p}} = *'((T_{-p} - T_{-\epsilon})w + 1) \text{ for } p > 1+\epsilon.$$

Next we treat the case $p = 1+\epsilon$. $\alpha+\beta$ equals $T_1 - T_{-1-\epsilon}$ and on E

$$w = (-1)\tilde{m}_{T_{-\epsilon}-T_{-\epsilon-1}} + (-1)(T_1 - T_{-1-\epsilon})\tilde{m}_{T_1-T_{-\epsilon-1}}$$

and

$$\tilde{m}_{T_{-\epsilon}-T_{-\epsilon-1}} = -\tilde{n}_{T_{-\epsilon}-T_{-\epsilon-1}} = 1/(T_{-1-\epsilon} - T_{-\epsilon}).$$

We find that $\tilde{m}_{T_1-T_{-1-\epsilon}} = *((T-_{1\epsilon} - T_{-\epsilon})w+1)$ and

$$\tilde{m}_{T_{1+\epsilon}-T_{-1-\epsilon}} = *'((T_{-1-\epsilon})w + 1).$$

If $p = \epsilon = 1$ then $\tilde{m}_\alpha = \tilde{m}_{T_1-T_{-1}}$ is constant. Finally we note that the factor $e(\alpha,\beta) = 1$ for our representation (5.6).

We are now in a position to carry out the transfer of the subregular germ. If $w_n = w'/(-Rw'+1), \sigma(w_n) = \delta w_n/(A_\sigma^n w_n+1), \sigma(w') = \delta w'/(A'_\sigma w'+1), \delta = \pm 1$, then this is defined over F provided

$$A_\sigma^n + \delta\sigma(R) - R = A'_\sigma.$$

We follow steps 2 and 3 of (5.5). For $w' = w_{2h}$ we let $R = 0$. For $w' = w_{n-2h}$ we let $R = T_{-p} - T_{-\epsilon}$. We verify that these maps are defined over F by using (5.6) and (5.8) and checking on generators. We may take $e(\alpha,\beta) = 1, e(\beta,\alpha) = -1$. $\delta = 1$ except for σ_0 where $\delta = -1$.

Generator	$\mathbf{A}_\sigma^n$	$\delta\sigma(\mathbf{R}) - \mathbf{R}$	$\mathbf{A}_\sigma^{2h}$
σ_0	$T_\epsilon - T_{-\epsilon}$	$(T_p - T_{-\epsilon}) - T_{-p} + T_{-\epsilon}$	$T_p - T_{-p}$
$\sigma_{T_1-T_{-\epsilon}}$	$T_1 - T_{-\epsilon}$	$T_{-\epsilon} - T_1$	0
$\sigma_{T_{-\epsilon}-T_{-\epsilon-1}}$	$T_{-\epsilon-1} - T_{-\epsilon}$	$T_{-\epsilon} - T_{-\epsilon-1}$	0
$\sigma_{T_p-T_{-p}}$	0	$T_p - T_{-p}$	$T_p - T_{-p}$
$\sigma_{T_{-p}-T_{-p-1}}$	0	$T_{-p-1} - T_{-p}$	$T_{-p-1} - T_{-p}$
others	0	0	0

Thus $A_\sigma^n + \delta\sigma(R) - R = A_\sigma^{2h}$.

Now we check that the map $w_n = w_{n-2h}$ is defined over F.

Generator	$\mathbf{A}_\sigma^n$	$\delta\sigma(\mathbf{R}) - \mathbf{R}$	$\mathbf{A}_\sigma^{n-2h}$
σ_0	$T_\epsilon - T_{-\epsilon}$	0	$T_\epsilon - T_{-\epsilon}$
$\sigma_{T_1-T_{-\epsilon}}$	$T_1 - T_{-\epsilon}$	0	$T_{-\epsilon} - T_1$
$\sigma_{T_{-\epsilon}-T_{-\epsilon-1}}$	$T_{-\epsilon-1} - T_{-\epsilon}$	0	$T_{-\epsilon} - T_{-\epsilon-1}$
$\sigma_{T_p-T_{-p}}$	0 (5.8)	0	0
others	0	0	0

Thus $A^n_\sigma + \delta\sigma(R) - R = A^{n-2h}_\sigma$.

We check that the cocycles are carried into cocycles for the transfer to $U(2h)$. We use (5.6), (5.15), (*I*.5.3) and (*I*.5.7). Note that $\zeta = 1$ for quasi-split groups.

Generator	$\mathbf{b}_{n,2h}$	$\mathbf{a}_{2h}$
σ_0	1	1
$\sigma_{T_p - T_{-p}}$	$((T_p - T_{-\epsilon})w_n + 1)/((T_{-p} - T_{-\epsilon})w_n + 1)$ $= (T_p - T_{-p})w_{2h} + 1$	$(T_p - T_{-p})w_{2h} + 1$
others	1	1

Thus the cocycles correspond on the transfer to $U(2h)$.

Finally we must check that the cocycles correspond on the transfer to $U(n - 2h)$.

Generator	$\mathbf{b}_{n,2h}$	$\mathbf{a}_n$	$\mathbf{a}_{n-2h}$
σ_0	1	$1/\zeta$	1
	1	$w/x(\gamma)$	$w/x(\gamma)$
$\sigma_{T_1 - T_{-\epsilon}}$	1	$(T_1 - T_{-\epsilon})w_n + 1$	$(T_1 - T_{-\epsilon})w_{n-2h} + 1$
$\sigma_{T_p - T_{-p}}$	x	x	1
others	1	1	1

where $x = ((T_p - T_{-\epsilon})w_n + 1)/((T_{-p} - T_{-\epsilon})w_n + 1)$. So $b^{-1}_{n,2h} a_n = \zeta^{-1} \cdot a_{n-2h}$ where we identify ζ with a cocycle in the obvious way. In the degenerate case $k = h$, $n - 2h = 1$. $w = w_{n-2h}$ is defined over F (we may exclude the generators $\sigma_{T_1 - T_0}$ and $\sigma_{T_0 - T_{-1}}$) and $\eta_K(a_{n-2h}) = \eta_K(-w/x(\gamma))$. The principal value integral $\int \eta_K(w)|dw/w^2|$ is zero [16]. So everything checks out. This completes the proof of the transfer.

REMARK. We make a few observations about the transfer. First the arguments are independent of the Cartan subgroup. Second the singularities of the variety Y_1 ultimately played no role in the expression for the subregular germ. Finally we note that the transfer factor enters into the transfer of the subregular germ in almost a trivial way.

REFERENCES

1. J. Arthur, *On some problems suggested by the Trace Formula.*, Lecture Notes in Math., vol. 1041, Springer-Verlag, Berlin, 1984.
2. A. Borel, *Automorphic L-functions*, Automorphic Forms, Representations and L-functions, Proceedings of Symposia in Pure Mathematics, Volume XXXIII, Part 2, American Mathematical Society, Providence, Rhode Island, 1979.
3. R.W. Carter, *Finite Groups of Lie Type: Conjugacy Classes and Complex Characters*, Wiley-Interscience, John Wiley and Sons, Chichester, 1985.
4. M. Golubitsky and B. Rothschild, *Primitive Subalgebras of Exceptional Lie Algebras*, Pacific Journal of Mathematics **39 No. 2** (1971).
5. Harish-Chandra, *Admissible invariant distributions on reductive p-adic groups*, Queen's Papers in Pure and Applied Math. **48** (1978), 281–347.
6. Harish-Chandra, *Harmonic Analysis on Reductive p-adic Groups*, Notes by G. van Dijk, Springer-Verlag, Berlin, 1970.
7. R. Hartshorne, *Algebraic Geometry*, Springer-Verlag, New York, 1977.
8. H. Hironaka, *Algebraic Geometry*, J.J. Sylvester Sympos. Johns Hopkins Univ. 1976, Johns Hopkins Univ. Press, Baltimore, Md., 1977, pp. 52–125..
9. H. Hironaka, *Memorias de Matematica del Instituto "Jorge Juan"*, vol. 29, Consejo Superior de Investigaciones Cientificas, Madrid, 1975.
10. J. E. Humphreys, *Linear Algebraic Groups*, Springer-Verlag, New York, 1981.
11. J.-I. Igusa, *Lectures on forms of higher degree*, Tata Institute of Fundamental Research, Bombay, 1978.
12. R. Kottwitz, *Rational Conjugacy Classes in Reductive Groups*, Duke Math. Journal **49, No. 4** (1982), 785–806.
13. M. Kneser, *Galois-Kohomologie halbeinfacher algebraischer Gruppen über p-adischen Körpern*, Math. Z. **89** (1965), 250–272.
14. R. Langlands, *Les debuts d'une formule des traces stable*, vol. 13, Publications math. de l'Univ. de Paris VII, 1983.
15. R. Langlands, *Orbital Integrals on Forms of* $SL(3)$, *I*, American Journal of Mathematics **105** (1983), 465–506.
16. R. Langlands and D. Shelstad, *On Principal Values on p-adic Manifolds*, Lie Group Representations II, Lecture Notes in Math., vol. 1041, Springer-Verlag, Berlin, 1984.
17. R. Langlands and D. Shelstad, *On the Definition of the Transfer Factors*, Math. Ann. **278** (1987), 219–271.
18. D. Mumford, *The Red Book of Varieties and Schemes*, Lecture Notes in Math., vol. 1358, Springer-Verlag, Berlin, 1988.

19. J.-P. Serre, *Corps Locaux*, Actualités scientifiques et industrielles, vol. 1296, Hermann, Paris, 1962.
20. J.-P. Serre, *Cohomologie galoisienne*, Lecture Notes in Math., vol. 5, Springer-Verlag, Berlin, 1964.
21. P. Slodowy, *Simple Singularities and Simple Algebraic Groups*, Lecture Notes in Math., vol. 815, Springer-Verlag, Berlin, 1980.
22. N. Spaltenstein, *Classes Unipotentes et Sous-groupes de Borel*, Lecture Notes in Math., vol. 946., Springer-Verlag, Berlin, 1982.
23. R. Steinberg, *Conjugacy Classes in Algebraic Groups*, Lecture Notes in Math., vol. 366., Springer-Verlag, Berlin, 1974.

DEPARTMENT OF MATHEMATICS, UNIVERSITY OF CHICAGO, CHICAGO ILLINOIS, 60637
E-mail: hales@zaphod.uchicago.edu

LIST OF NOTATION AND CONVENTIONS

α	I.2	simple root
α	IV.7	root next in size to β in S_-
α-cell	IV.5	
α-wall	IV.4	
α-chamber	IV.4	
α	II.7	short simple root of G_2
α	III.1	positive simple root in a rank two group
α^v	I.5	coroot associated to α
β		positive simple root
β	II.7	long simple root of G_2
β	III.1	positive simple root in a rank two group
β	IV.7	largest simple root of S_-
$\beta(E_\Sigma)$	II.9	Igusa constants
β	V.3	Igusa constant, see Langlands [15]
γ	III.1	positive root in a rank two group
γ	II.7	root $\alpha+\beta$ of G_2
γ_j	VII.5.13	root $T_j - T_{-j}$
Γ	I.1,I.6	regular curve in T
Γ	I.6	$\Gamma \setminus \{0\}$
$\Gamma_O, \Gamma_{O'}$	VII.4	subregular germs
Γ_0	IV.3.2	a graph
Γ_1	IV.4	minimal tree in Γ_0 containing extremal $\alpha, \beta \in S_-$
Γ'	IV.4	pruned Γ_1, as a graph it equals S_-
δ	III.1	positive root in a rank two group
δ	II.7	root $2\alpha+\beta$ of G_2
δ	V.2.3	$\delta(\xi) = x(\beta)/\xi x(\gamma)$
Δ_Γ	VI.2	$= m_\kappa(e')$, $e' = e(T(W,\alpha)$: α simple
Δ'	VI.2	representatives of orbits of simple roots under $\mathrm{Gal}(\bar{F}/F)$

ϵ	III.1	positive root in a rank two group
$\epsilon_{\pm\alpha}(x)$	I.5	$\exp(x\,X_{\pm\alpha})$, $X_{\pm\alpha}$ root vectors
ϵ	II.7	root $3\alpha+\beta$ of G_2
ϵ	VII.5.5	$\epsilon=1$ for n even and $\epsilon=0$ for n odd
ϵ_j	VII.5.5	permutation $(j,-j)$ of $-k,\dots,k$
ζ	III.1	positive root in a rank two group
ζ	II.7	root $3\alpha+2\beta$ of G_2
η_K	V.4,VII.4	nontrivial character of $H^1(\mathrm{Gal}(\bar F/F),U(1))$
η_K		nontrivial quadratic character of $F^\times$
κ	I.1,I.5	character on $H^1(\mathrm{Gal}(\bar F/F),T)$
κ^α	VI.2	character of $F_\alpha^\times$
$\kappa(E)$	V.2	character such that $f/\kappa(E)(\mu)$ extends generically to E
$\kappa(E)$	VI.2	$\prod(\kappa^\alpha)^{e(\alpha)}$, $e(\alpha)$ an integral multiplicity
$\kappa(E_0)$	VI.2	$=\kappa_0$
κ_0	VI.2	$\prod\kappa^\alpha$
λ	I.1	local parameter on Γ at p
λ	II.4	pullback of a local parameter on Γ to Y
ν	I.2,II.1	an element of N_∞
ν_2	V.3	form on $E_\alpha\cap E_\beta$
ξ	I.5,V.1.2	coordinate in $N_{qs\infty}/N_\alpha$ on $U(\alpha,\beta)\cap E_\alpha(u)^0$
ξ	IV.5	a wall of an α-chamber
$\xi: X_1\to T$	I.6	a morphism
$\xi: {}^LH\to{}^LG$		an embedding of L-groups
$\pi_1: X_1\to G$	I.6	a morphism
$\pi_1: Y_1\to G$	I.5	a morphism
ρ	VII.1	homomorphism $\mathrm{Gal}(\bar F/F)\to\mathrm{Outer}({}^LH^0)$
σ_α	I.5,II.6,V.5.2	simple reflection in the Weyl group
σ_ω	II.6	representative of $\omega\in\Omega$ in $N_G(T_0)$
σ_*	VI.1	action of Galois group on simple roots in a quasisplit group
σ_{sp}	VI.1	action of the Galois group in G_{sp}

σ_0	VI.1	element of $\tilde{\Omega}$
σ_T	III.1	permutation of Weyl chambers associated to $\sigma \in \mathrm{Gal}(\bar{F}/F)$ and T
Σ	VI.2	an element of Δ'
$\phi : T^0 \times T\backslash G \to X_1$	I.6	a morphism
ω_α	V.5.2	$\omega_\alpha \in N_G(T_0)$
ω_σ	VI.1	element of $N_{P_\alpha}(T_0)_{adj}$
ω_Z	I.2,V.2	invariant form of top degree on Z, $Z = M, T, T\backslash G, X$ or Y
Ω	I.2	Weyl group of G with respect to T
$\tilde{\Omega}$	VI.1	the extended Weyl group or direct product $\Omega \times \mathbb{Z}/2\mathbb{Z}$
Ω'	III.1	Image($\mathrm{Gal}(\bar{F}/F) \to \tilde{\Omega}$), $\tilde{\Omega}$ extended Weyl group
adjacent walls	IV.4	
adjoint conjugacy	I.6	a conjugacy class in the adjoint group
a_σ	V.4	cocycle in $H^1(\mathrm{Gal}(\bar{F}/F), U(1))$
a_σ	VI.1	automorphism of $(G_{sp}, B_{sp}, T_{sp}, \{X_\gamma\})$
$a_\sigma(w)$	VI.2.3	a cocycle
a_n	VII.5.5	the cocycle a for the group $U(n)$
$a(E_\Sigma)$	II.9	Igusa constants
A_σ	I.5	a cocycle of $Z^1(\mathrm{Gal}(\bar{F}/F), N_{G_{qs}}(T_{qs})_{ad})$
$A(X), A_r(X)$	V.3	Igusa data, see Langlands [15]
A_n	I.3	group or algebra of type A_n
$\mathbb{A}^r$		affine r-space
big nodes	IV.4	
big chamber	IV.4	
big wall	IV.4	
b	II.1	element of B_0
$\mathbf{b} = b_{n,2h}$	VII.5.5	a cocycle in $H^1(U(1))$
$\mathbb{B}$	I.2	Borel subgroup containing T
$\mathbb{B}(W)$	I.2	$\mathbb{B}^\omega$, $W = W(\omega)$.
$(B(W))$	I.2	star in S
B_0, B_∞	I.2	a pair of opposite Borel subgroups
B_{qs}	I.5	Borel subgroup over F containing T_{qs}
B_{sp}	VI.1	Borel subgroup in split form
B_+	V.5	intersection of lines ℓ_α, ℓ_β
B_-	V.5.1	intersection of lines ℓ_α, ℓ'_α
B.I,...,B.IV	III.1	zero patterns
B_n	I.3	group or algebra of type B_n

$\mathbb{C}$		complex numbers
$C_G(x)$		centralizer in G of x
$C_G(x)_{red}$	VII.5	the reductive part of the centralizer of x
C_n	I.3	group or algebra of type C_n
divisor	I.6	
– fundamental	II.9	
– O	I.6	a divisor over the unipotent class O
– regular	I.6	a divisor over a regular unipotent class
– spurious	II.9	
– subregular	I.6	a divisor over a subregular unipotent class
D_n	I.3	group or algebra of type D_n
Dynkin curve	IV.1,VI.1	$(B\backslash G)_u$
external wall	IV.5	
e	I.5,II.1	a star, element of S
$e(p)$	VI.2	a star depending on a point p
$e = e(\alpha_1, \alpha_2)$	V.6	constant defined by $\exp(X_{-\alpha_1})\exp(X_{\alpha_1+\alpha_2})\exp(-X_{-\alpha_1}) = \dots$
e_ℓ	VII.5.6	constants in F
E	I.6	a divisor
$E(u)$	I.6	the fibre in E over $u \in G$
E_α	V.1	a subregular divisor, α-simple, $E_\alpha = E_\Sigma$, $\Sigma = \{\alpha\}$
$E_\alpha(u)$	V.1.1	$\pi^{-1}(u) \cap E_\alpha$, fibre over u in E_α
$E_\alpha(u)^0$	V.1.1	an irreducible component of $E_\alpha(u)$
$E(\ell_\alpha, u)$	V.1.2	a component of $E_\alpha(u)$ indexed by the line ℓ_α
E_0	V.1	the regular divisor, $E_0 = E_\Sigma$, $\Sigma = \emptyset$
E	VII.5.14	$E_{\alpha_k} \cap E_{\alpha_{k+1}}$ or E_{α_k}
E_n	I.3	group or algebra of type E_n, $n = 6, 7, 8$
fundamental		
– α-cell	IV.5	
– divisor	II.9	
f	I.5	a locally constant fuunction of compact support on G
F	I.	p-adic field of characteristic 0.
$\bar{F}$	I.	algebraic closure of F
F_α	VI.2	field extension of F fixing α
F_4	I.3	group or algebra of type F_4
$F_1(\beta, \theta, f)$	V.3	a term of the asymptotic expansion, see Langlands [15]
G	I.	reductive group defined over F

G_{adj}	I.6	the adjoint group of G
G_{der}	II.5	the derived group of G
G_{in}	VII.4	inner form of G
G_{in}	I.5	an inner form of G_{qs}
G_{qs}	I.5	a quasi-split group, a form of G
G_{sp}	VI.1	split form of G
G_2	I.3	group or algebra of type G_2
h_2	V.3	function on $E_\alpha \cap E_\beta$
$H^1(T)$	I.5	$H^1(T) = H^1(\mathrm{Gal}(\bar{F}/F), T)$ first cohomology group with coefficients in T
H	I.5,VII.1	endoscopic group
H	VII.5	the endoscopic group $U(n-2h) \times U(2h)$ of $U(n)$
$\underline{H}$	VII.5.11	subgroup $U(2h) \times U(n-2h)$ of $G = U(n)$
H'	VII.5.11	subgroup of $U(n)$ conjugate to $\underline{H}$
K		field extension of F
K_X	V.2.3	field of rational functions on $Lie(T)$
K_α	IV.7	an α-cell in Z_α^-
ℓ_α	V.1.2	a projective line of type α in the flag variety
$L_p(W)$	IV.1	a set of simple roots
$L^- = L_{\beta_i}^-$	IV.5,IV.7	
m_α	II.2	$\beta = \sum m(\alpha)\alpha$
$m_\kappa(e)$	I.5	the function $g \to \kappa(\sigma(g)g^{-1})$, $\sigma \in \mathrm{Gal}(\bar{F}/F)$, $g \in (T\backslash G)(F)$
M	I.2	Springer variety $B \times_B G$
M	V.3	constant $m + m_1 + m_2 + am(\mu_3)$
node	I.3	a node is an $\{i,j\}$-residue in the Coxeter complex of Ω
– big	IV.4	
– solid	IV.3	
– special	III.1	
n_w	II.1	a coordinate in N_∞: $B_0^{n_w} = B(W)$
$N_G(T)$		normalizer in G of T
N	I.2	unipotent radical of $\mathbb{B}$
N_∞	I.2	the unipotent radical of B_∞
N_{qs}	I.5	unipotent radical of B_{qs}
$N_{qs\infty}$	I.5	unipotent radical of Borel opposite B_{qs} through T_{qs}

N_α	I.5	radical of the parabolic subgroup $B^{op}_{qs}\langle\sigma_\alpha\rangle B^{op}_{qs}$, $op =$ opposite
N_ω	II.6	subgroup of N_0
N	VI.2.3	the norm map from K to F, K a field
obtusely adjacent	IV.4	
O-divisor	I.6	a divisor over the unipotent class O
O	II.9	unipotent class O
proximate chamber	IV.4	
p	I.1	point on a curve Γ
p	II.1	a point in X_1
p	I.	residue characteristic of F
$\mathbb{P}_\alpha$	II.6	parabolic subgroup containing B_0
P_α	IV.1,V.1.2	a parabolic subgroup of type α containing B_0
$\mathbb{P}^1$	VII.4	projective line
Q-chamber	IV.5	
regular divisor	I.6	a divisor over a regular unipotent class
r	I.2	the semisimple rank of G
R	IV.3	roots α such that $z(W,\alpha) \neq 0$ for some W
R_ω	II.7	$\{\beta > 0 \mid \omega\beta < 0\}$
R_1, R_2	VII.5.5	elements of K_X
solid node	IV.3	
special node	III.1	
spurious divisor	II.9	
subregular		
– divisor	I.6	a divisor over a subregular unipotent class
– conjugacy class	IV.1	
– unipotent	III.3	
S	I.2	variety of stars
S'	I.2	subvariety of S

S''	I.2	open subvariety of S_1
S^0	I.2	variety of regular stars
S_1	I.2	first resolution of S
$S(B_\infty)$	I.2	a coordinate patch on S
$S(B_\infty, B_0)$	I.2	a coordinate patch
S_2	IV.4	
S	IV.3	roots α such that $z_1(W,\alpha)=0$ for some W
S_-	IV.4	$S_- = S$ except for D_4, $\lvert S\rvert \geq 4, \ldots$
type of a vertex	IV.3.2	
t_σ	I.5	a cocycle on $T(R)$, R the field of rational functions on S
T	I.2	Cartan subgroup in G over F
T^0	I.2	regular elements of T
T_0	I.2	intersection of opposite Borel subgroups B_0, B_∞
T_0	V.5.2	Cartan subgroup in $B_+ \cap B_-$
T_{in}	I.5	Cartan subgroup in G_{in}
T_{qs}	I.5	maximally split Cartan subgroup of G_{qs}
T_{sp}	VI.1	Cartan subgroup in split form
$T(W,\alpha)$	IV.1	$(1-\gamma^{-1})/\lambda$
T_σ	I.5	a twisted cocycle in $Z^1(T_{qs})$
T_i	VII.5.5	the ith character on the Cartan subalgebra
$U(1)$	V.4	unitary group in 1-variable
$U(\alpha',\alpha'')$	V.1.1	a coordinate patch in Y_s
U_1, U_2	V.3	coordinate patches
V^n	I.2	$n = \lvert\Omega\rvert$-fold product of the flag variety
$w = w(\alpha+\beta)$	V.2	a coordinate on $U(\alpha,\beta)\cap E_\alpha(u)^0$
w_n	VII.5.5	the variable w for the group $U(n)$
W_+	I.2	positive Weyl chamber for $\mathbb{B}$
$W(\omega)$	I.2	$\omega^{-1}W_+$, $\omega \in \Omega$
$\mathbb{W}$	IV.4	a big chamber
$\mathbf{W}$		a Q-chamber
x_β	II.1	$x_\beta(b)$ is the βth coefficient of $b \in B_0$
$x(W,\beta)$	II.1	a local coordinate
$(x_{-k}, \ldots, x_k)$	VII.5.5	element of a Cartan subalgebra for $U(n)$, $2k+1-\epsilon = n$
X	I.2	closure of X^0 in $G \times S$

X'	I.2	closure of X^0 in $G \times S'$
X''	I.2	closure of X^0 in $G \times S''$
$X(B_\infty, B_0)$	I.2	a coordinate patch on X
X^0	I.2	subvariety of $G \times S^0$
X_1	I.2	closure of X^0 in $G \times S_1$
$X_{\pm\alpha}$	I.2,VI.2	root vectors
$X_{\pm\alpha}, H_\alpha$	II.1	a Lie triple
$X(H)$	VII.3	a set of characters $\theta : F^\times \to \mathbb{C}^\times$

y_β	II.2	βth coefficient of t^n
Y	I.2	closure of Y^0 in X
Y'	I.2	closure of Y^0 in X'
Y''	I.2	closure of Y^0 in X''
$Y(B_\infty, B_0)$	I.2	a coordinate patch on Y
Y_Γ	I.2	resolution of singularities of Y_1
Y_Γ	I.6	resolution of Y_1
Y^0	I.2	restriction of X^0 to Γ
Y_1	I.2	closure of Y^0 in X_1
Y_s	V.1	an open subvariety of Y''
$Y(T,\kappa)$	VII.3	a set of characters $\theta : F^\times \mathbb{C}^\times$

zero pattern	III.1	
$\hat{z}(\alpha)$	IV.4	a variable in S_2
$\hat{z}(W,\alpha)$	IV.4	a variable in S_2
$\tilde{z}(W,\alpha)$	IV.1	$\tilde{z}(W,\alpha) = z(W,\alpha)/z(W_\alpha,\alpha)$
$z(W,\alpha)$	I.2	a coordinate on $S(B_\infty, B_0)$
$z_1(W,\alpha)$	I.2	homogeneous coordinate in $S_1(B_\infty, B_0)$
$z_1(W,\alpha)$	II.2	$z_1(W,\alpha) = z(W,\alpha)/z(\alpha)$
$\mathbb{Z}$		ring of integers
Z	VII.5	center of G
Z_α^-	IV.7	union of α-chambers
Z_β	IV.7	union of β-chambers

Editorial Information

To be published in the *Memoirs*, a paper must be correct, new, nontrivial, and significant. Further, it must be well written and of interest to a substantial number of mathematicians. Piecemeal results, such as an inconclusive step toward an unproved major theorem or a minor variation on a known result, are in general not acceptable for publication. *Transactions* Editors shall solicit and encourage publication of worthy papers. Papers appearing in *Memoirs* are generally longer than those appearing in *Transactions* with which it shares an editorial committee.

As of July 1, 1992, the backlog for this journal was approximately 8 volumes. This estimate is the result of dividing the number of manuscripts for this journal in the Providence office that have not yet gone to the printer on the above date by the average number of monographs per volume over the previous twelve months. (There are 6 volumes per year, each containing about 3 or 4 numbers.)

A Copyright Transfer Agreement is required before a paper will be published in this journal. By submitting a paper to this journal, authors certify that the manuscript has not been submitted to nor is it under consideration for publication by another journal, conference proceedings, or similar publication.

Information for Authors

Memoirs are printed by photo-offset from camera copy fully prepared by the author. This means that the finished book will look exactly like the copy submitted.

The paper must contain a *descriptive title* and an *abstract* that summarizes the article in language suitable for workers in the general field (algebra, analysis, etc.). The *descriptive title* should be short, but informative; useless or vague phrases such as "some remarks about" or "concerning" should be avoided. The *abstract* should be at least one complete sentence, and at most 300 words. Included with the footnotes to the paper, there should be the 1991 *Mathematics Subject Classification* representing the primary and secondary subjects of the article. This may be followed by a list of *key words and phrases* describing the subject matter of the article and taken from it. A list of the numbers may be found in the annual index of *Mathematical Reviews*, published with the December issue starting in 1990, as well as from the electronic service e-MATH [**telnet e-MATH.ams.com** (or **telnet 130.44.1.100**). Login and password are **e-math**]. For journal abbreviations used in bibliographies, see the list of serials in the latest *Mathematical Reviews* annual index. When the manuscript is submitted, authors should supply the editor with electronic addresses if available. These will be printed after the postal address at the end of each article.

Electronically-prepared manuscripts. The AMS encourages submission of electronically-prepared manuscripts in $\mathcal{A}\mathcal{M}\mathcal{S}$-TEX or $\mathcal{A}\mathcal{M}\mathcal{S}$-LATEX. To this end, the Society has prepared "preprint" style files, specifically the amsppt style of $\mathcal{A}\mathcal{M}\mathcal{S}$-TEX and the amsart style of $\mathcal{A}\mathcal{M}\mathcal{S}$-LATEX, which will simplify the work of authors and of the production staff. Those authors who make use of these style files from the beginning of the writing process will further reduce their own effort.

Guidelines for Preparing Electronic Manuscripts provide additional assistance and are available for use with either $\mathcal{A}\mathcal{M}\mathcal{S}$-TEX or $\mathcal{A}\mathcal{M}\mathcal{S}$-LATEX. Authors with FTP access may obtain these *Guidelines* from the Society's Internet node e-MATH.ams.com (130.44.1.100). For those without FTP access they can be obtained free of charge from the e-mail address guide-elec@math.ams.com (Internet) or from the Publications Department, P. O. Box 6248, Providence, RI 02940-6248. When requesting *Guidelines* please specify which version you want.

Electronic manuscripts should be sent to the Providence office only after the paper has been accepted for publication. Please send electronically prepared manuscript files via e-mail to pub-submit@math.ams.com (Internet) or on diskettes to the Publications Department address listed above. When submitting electronic manuscripts please be sure to include a message indicating in which publication the paper has been accepted.

For papers not prepared electronically, model paper may be obtained free of charge from the Editorial Department at the address below.

Two copies of the paper should be sent directly to the appropriate Editor and the author should keep one copy. At that time authors should indicate if the paper has been prepared using $\mathcal{A}\mathcal{M}\mathcal{S}$-TEX or $\mathcal{A}\mathcal{M}\mathcal{S}$-LATEX. The *Guide for Authors of Memoirs* gives detailed information on preparing papers for *Memoirs* and may be obtained free of charge from AMS, Editorial Department, P. O. Box 6248, Providence, RI 02940-6248. The *Manual for Authors of Mathematical Papers* should be consulted for symbols and style conventions. The *Manual* may be obtained free of charge from the e-mail address cust-serv@math.ams.com or from the Customer Services Department, at the address above.

Any inquiries concerning a paper that has been accepted for publication should be sent directly to the Editorial Department, American Mathematical Society, P. O. Box 6248, Providence, RI 02940-6248.

Editors

Recent Titles in This Series

(Continued from the front of this publication)

443 **D. J. Benson and F. R. Cohen,** Mapping class groups of low genus and their cohomology, 1991

442 **Rodolfo H. Torres,** Boundedness results for operators with singular kernels on distribution spaces, 1991

441 **Gary M. Seitz,** Maximal subgroups of exceptional algebraic groups, 1991

440 **Bjorn Jawerth and Mario Milman,** Extrapolation theory with applications, 1991

439 **Brian Parshall and Jian-pan Wang,** Quantum linear groups, 1991

438 **Angelo Felice Lopez,** Noether-Lefschetz theory and the Picard group of projective surfaces, 1991

437 **Dennis A. Hejhal,** Regular b-groups, degenerating Riemann surfaces, and spectral theory, 1990

436 **J. E. Marsden, R. Montgomery, and T. Ratiu,** Reduction, symmetry, and phase mechanics, 1990

435 **Aloys Krieg,** Hecke algebras, 1990

434 **François Treves,** Homotopy formulas in the tangential Cauchy-Riemann complex, 1990

433 **Boris Youssin,** Newton polyhedra without coordinates Newton polyhedra of ideals, 1990

432 **M. W. Liebeck, C. E. Praeger, and J. Saxl,** The maximal factorizations of the finite simple groups and their automorphism groups, 1990

431 **Thomas G. Goodwillie,** A multiple disjunction lemma for smooth concordance embeddings, 1990

430 **G. M. Benkart, D. J. Britten, and F. W. Lemire,** Stability in modules for classical Lie algebras: A constructive approach, 1990

429 **Etsuko Bannai,** Positive definite unimodular lattices with trivial automorphism groups, 1990

428 **Loren N. Argabright and Jesús Gil de Lamadrid,** Almost periodic measures, 1990

427 **Tim D. Cochran,** Derivatives of links: Milnor's concordance invariants and Massey's products, 1990

426 **Jan Mycielski, Pavel Pudlák, and Alan S. Stern,** A lattice of chapters of mathematics: Interpretations between theorems, 1990

425 **Wiesław Pawłucki,** Points de Nash des ensembles sous-analytiques, 1990

424 **Yasuo Watatani,** Index for C^*-subalgebras, 1990

423 **Shek-Tung Wong,** The meromorphic continuation and functional equations of cuspidal Eisenstein series for maximal cuspidal subgroups, 1990

422 **A. Hinkkanen,** The structure of certain quasisymmetric groups, 1990

421 **H. W. Broer, G. B. Huitema, F. Takens, and B. L. J. Braaksma,** Unfoldings and bifurcations of quasi-periodic tori, 1990

420 **Tom Lindstrøm,** Brownian motion on nested fractals, 1990

419 **Ronald S. Irving,** A filtered category $\mathcal{O}_S$ and applications, 1989

418 **Hiroaki Hijikata, Arnold K. Pizer, and Thomas R. Shemanske,** The basis problem for modular forms on $\Gamma_0(N)$, 1989

417 **John B. Conway, Domingo A. Herrero, and Bernard B. Morrel,** Completing the Riesz-Dunford functional calculus, 1989

416 **Jacek Nikiel,** Topologies on pseudo-trees and applications, 1989

415 **Amnon Neeman,** Ueda theory: Theorems and problems, 1989

(See the AMS catalog for earlier titles)